Books should be returned on or before the
last date sta

D0297414

1 1 MAR 2005

- 5 AUG 2006

1 9 DEC 2006 2 7 NOV 2012
              - 5 JUL 2013
              1 6 AUG 2013

HEADQUARTERS

HEADQUARTER 1 9 MAR 2015

1 4 SEP 2009   - 6 MAY 2015

< 7 MAR 2017

2 5 SEP 2017

1 1 NOV 2017
- 7 DEC 2017

1 9 JUL 2018
1 5 SEP 2018

- 2 SEP 2015

- 9 JAN 2016

The Sun

The Sun

A BIOGRAPHY

David Whitehouse

WILEY

Published in 2005 by   John Wiley & Sons, Ltd, The Atrium, Southern Gate
Chichester, West Sussex, PO19 8SQ, England
Phone   (+44) 1243 779777

Copyright © 2004   David Whitehouse

Email (for orders and customer service enquiries): cs-books@wiley.co.uk
Visit our Home Page on www.wiley.co.uk or www.wiley.com

David Whitehouse   Whitehouse, Dav   t, Designs and Patents Act 1988, to be identif

Other Wiley Edit The Sun : a

John Wiley & Sons   biography /   30, USA
David
Jossey-Bass, 989 M   523.   741, USA

Wiley-VCH Verlag   7   , Germany

John Wiley & Sons   1591651   ensland, 4064, Australia

John Wiley & Sons (Asia) Pte Ltd, 2 Clementi Loop #02-01, Jin Xing Distripark, Singapore 129809

John Wiley & Sons Canada Ltd, 22 Worcester Road, Etobicoke, Ontario, Canada, M9W 1L1

Wiley also publishes its books in a variety of electronic formats. Some content that appears in print may not be available in electronic books.

British Library Cataloguing in Publication Data
A catalogue record for this book is available from the British Library

ISBN 0-470-09296-3

Typeset in 10½/13½ Photina by Mathematical Composition Setters Ltd, Salisbury, Wiltshire
Printed and bound in Great Britain by T.J. International Ltd, Padstow, Cornwall

10 9 8 7 6 5 4 3 2 1

To Jill, Christopher, Lucy and Emily, with thanks for all the sunny days in the past, and for those to come.

Contents

Acknowledgements

When I wrote my first book, *The Moon: A Biography* in 2001, I knew that someday I would have to write a corresponding book about the Sun, and I also knew it would be a bit more difficult to put together.

The Moon is a ball of rock – though not just a ball of rock – and its days of activity are over. It is a passive object, whereas the Sun is not. The Sun is almost alive in its production of energy and its cycles and rhythms. It is one of the fundamental building blocks of the cosmos and has a much higher place in the ecology of the Universe than the Moon does. The Moon died a long time ago, its story is one of ancient times and of past landscapes. The Sun's story is of the future. Its future and our future are intertwined.

When I started writing this book, in November 2003, the most energetic series of explosions ever seen on the surface of the Sun erupted. I do not know if this was a sign of approval or disapproval, but I took it as a helpful gift of topicality. I knew that the Sun was full of surprises and mysteries, but I did not realise how much.

When I was young and lived in Birmingham in the English Midlands, I used to carry my small telescope into the garden and set it on a wall. I remember cold November days with crisp snow on the ground and steely blue skies. During the day it was the Sun that attracted me, and the Moon during the night. I still remember family, friends and neighbours coming up to me for a look at the tiny black dots that were sunspots, shaking on a cardboard screen made from a cutout breakfast cereal packet. This book is for them, to say thank you.

I never got on with the Sun when I was a professional astronomer, due to my interests and the way my academic institution was organised. But it has always been there and as a science communicator I once more turned to it.

I would like to thank my agent, Lavinia Trevor, for her unswerving support and Sally Smith of John Wiley for her patience and expertise.

I would also like to publicly pay a debt I owe to Patrick Moore. It was he who inspired and nurtured a young boy's interest in astronomy with his books and television appearances. How I miss those Sunday nights when I would stay up late (well, to 11 o'clock) to watch him, in black and white, on *The Sky at Night*. Each month there was a new window on the Universe. I remember especially enjoying Patrick's trip underground to the mine in the US that was searching for neutrinos from the Sun, and his visit to W. M. Baxter's observatory, which was dedicated to the Sun. From then on I wanted an observatory like that. Thank you, Patrick.

Of the many people who have helped me through, none have done so more than my family. Without the support of my wife Jill and my children Christopher, Lucy and Emily, this book would not have been written. They endured the long nights when the Sun had gone down that I spent writing about it, and my confusion and irritation when the book's structure was not right. May the Sun always shine on them.

David Whitehouse
Hampshire, UK
August 2004

CHAPTER ONE

Bits of a star gone wrong

The President of the United States was missing. On the morning of 24 April 1984 Ronald Reagan was onboard *Air Force One* over the mid-Pacific, having left Hickam Air Force Base in Honolulu, en route to China via Guam, for a meeting with Premier Zhao. In his suite he was talking with his aides in Washington when the line suddenly went dead. The pilot told him that all communication channels between *Air Force One* and the outside world had been cut off. The most powerful man in the world was out of touch with the infrastructure of his government for over an hour.

The Soviets were also experiencing communication difficulties; they were not to blame. The pilots of the Presidential flight were experienced enough to know what was happening. It was the Sun, 150 million km away, that was responsible. There was a string of sunspots spread over 280,000 km across the Sun's surface – over 20 times the diameter of the Earth. Active Region 4474, as it had been designated, had been scrutinised by astronomers for days. Many were observing it: the Solar Tower telescope at Mt Wilson Observatory in California, the National Solar Observatories at Kitt Peak in Arizona and Sacramento Peak in New Mexico, radio telescopes scattered across the world, satellites in orbit and thousands of amateur astronomers from their back gardens. The region was crackling with electric and magnetic energy and had produced vast explosions called flares incorporating several powerful bursts of radiation as well as a very rare 'white light' flare and the most powerful X-ray-emitting flare ever seen up to then. Although the Sun had been declining in its surface activity for several years, as part of its regular 11-year solar cycle, there had been a dramatic recent surge in activity.

As astronomers watched, Active Region 4474 produced an explosion equivalent to billions of megaton h-bombs that heated gas in the Sun's atmosphere to tens of millions of degrees, throwing a trillion kilograms of it out into space. Shock waves from the blast rippled between the planets. The ejected gas cloud, with its pent-up magnetic energy, reached the Earth and triggered a so-called geomagnetic storm as its magnetic field caused the Earth's field to twist and writhe, bringing on radio communication blackouts. But such energies, titanic on a human scale, are but minor perturbations in the Sun's overall output, being only equal to the Sun's output for one hundredth of a second. Yet that energy, if it could be captured, would be enough to supply mankind's energy needs for 10,000 years. The Sun often shows us just how paltry is our desire and command of energy.

The flare from Active Region 4474 was detected by the orbiting *Solar Maximum Mission* satellite that had been circling the Earth since 1980, a year of peak solar activity. SMM carried a suite of state-of-the-art instruments to observe the Sun, in particular its flares and its total radiative output. Curiously, its first five years of data suggested that the Sun was dimming slowly, by a small but significant amount each year (0.02%). Some astronomers were puzzled, questioning whether the Sun would go on becoming dimmer or whether the dimming was part of a cycle to be followed by subtle brightening later. Measurements made with rockets and balloons also noted the decline and heightened the puzzle.

Earlier that month space shuttle mission 41-C had visited the SMM to repair it. Not long into its mission the SMM had experienced failures, so for the first time astronauts had grabbed a satellite and pulled it into the shuttle's cargo bay for a service before releasing it. Fortunately, the shuttle had landed on

April 13, so the problem of radiation from the Sun's outburst posing a danger for the crew was averted. Significantly, but not realised as such at the time, post-flight inspections showed that that particular shuttle had experienced what was described as 'enhanced primary O-ring erosion in the Right Hand Nozel Joint'. Two years later the same shuttle – *Challenger* – would be destroyed and its crew killed when an O-ring in one of its boosters failed and caused an explosion. The commander on that mission was Dick Scobee, who was also the pilot on the mission to rescue the SMM.

Mankind is here at the Sun's whim. Presidents are humbled before it and astronauts orbiting the earth seem puny in comparison and continually at its mercy. The Earth itself quivers and shakes in response to the forces the Sun can release: small changes in its activity cause the Earth to warm or chill, shift climatic zones, make deserts out of green lands and alter the fate of civilisations, perhaps including our own.

The sun is everywhere; in our past, present and future. But its light has given us more than life. It has given us a glimpse of how the Universe itself works. Not only is the Sun a fundamental constituent of the Universe, a star, but the very processes by which that star makes its light illuminate the workings of the cosmos at its most fundamental level.

We have had the Renaissance, the scientific and industrial revolutions. With our electric lights, televisions and fires, we think we do not rely on the Sun as much as we did generations ago. The mighty Sun has been removed from its prime position in our lives by our arrogance and by our indifference. But read on, and you will see that it will not be too long before the Sun once again reminds us that we are here because of it. There will come a time, perhaps within our lifetimes, when the Sun will no longer be our friend.

Science has made great strides in understanding our parent star, although there is still much about it that mystifies us. Nevertheless, one thing seems certain – if we survive, we will eventually have to do without it. The Sun is an average star, a typical building block of the Universe. However, because of this particular star's profound influence on the Earth, on our lives and on our future, it is essential that we understand it.

But where did the Sun come from? And how long will it last?

Only one God – the Sun.

CHAPTER TWO

The Goldilocks star

It begins again, as everyone knows it will – the event that has marked the start of each day. Ra has completed his journey through the underworld, faced the scorpion and returned triumphant. Inti has come back and looks to his people for human sacrifice. The all-knowing Nazambi surveys his world once more, dispassionately accounting for the souls that have departed since he left. The youthful and handsome Helios adjusts the long chiton of a charioteer, takes up the reins and pulls his four white horses to order. Pyrois, Eos, Aethon and Phlegon shake the dust from their manes and leap skyward from the shores of the great ocean Okeanos, watched by the immortals. Surya, of the 108 names, drives his five-horse chariot so that all can see his wisdom and benevolence.

Two great lights, so says the Book of Genesis: the greater light to rule the day, the lesser light to rule the night. All our days are ruled by the Sun. All cultures and civilisations have known it. In ancient times, whether it was the sole god or merely one of them, it was usually the most powerful. Is it a god? Certainly you cannot look at it. It was a god to the ancient Maya who called it Kinich Anau – the squint eyed. But hold your arm out and your thumb barely covers this god. We know that a million Earths could fit inside it. Yet it is still perhaps a god of sorts, for the Sun rules all of us. As well as the Earth, it is the only object in the Universe that we need. We could do without all the rest, all the other planets, stars and galaxies, but we need the Sun's presence bringing warmth and security in a cosmos that is mostly dark.

On the Côte d'Azur, or on Bondi and Malibu beaches, sun worshippers prepare to pay it homage. Farmers in Brazil tending the all-important maize crop hope that the coming day's sunlight will not be too harsh. In the morning mist, a grape is plucked from a vine in South Africa and tasted. The verdict of the experienced taster: it needs 'more sun'. Water evaporated from the Arabian Sea by the Sun's summer heat gathers in great clouds to be blown across the Indian subcontinent at the start of the monsoon season. Leaves start converting its light into carbohydrate. Ocean currents traverse the Earth powered by its rays, banishing the ice from shorelines as they redistribute the heat of the Sun to temperate climes. Across the globe the sunrise is the signal for work. As it reaches each section of the globe, vast armies of people respond to its command.

At its first rays the devoted bathe in the great religious rivers of southeast Asia and young girls nervously slip naked out of the water facing it in the hope of being blessed with a child. From research bases on the edge of Antarctica, from northern Canada and Scandanavia, people watch a shimmering green and red glow in the sky. Airline pilots adjust their course because of increased radiation. Engineers monitor the health of transcontinental oil pipelines. Controllers keep an eye on power relay systems and satellite operators watch nervously.

There is no such thing as a boring day on the Sun. Each day an army of astronomers and spacecraft study its ever-changing surface, returning images, colour coded by temperature, that show the birth and death of features in a realm where fundamental forces rule. There are wonders there: sunspots larger than the Earth, loops of superheated gas the height of ten Earths, unimaginable explosions, deep mysteries of matter.

The US National Solar Observatory at Sacramento Peak in New Mexico turns its huge Sun-reflecting mirror – a heliostat – so that it feeds the Sun's image down a huge long tube from which the air has been removed so that there will be no disturbing air currents to distort the image, around which astronomers sit wearing sunglasses. The McMath solar telescope at Kitt Peak, Arizona –

10

the world's largest solar telescope – watches the Sun rise, as does the Big Bear Solar Observatory near Pasadena, California that is located in the middle of a high-altitude lake, which provides a cooling and stabilising effect on the air flow around the telescope, and a better image.

The Nancay Radioheliograph in France turns its many radio dishes towards the Sun, scrutinising radiation from the hot gas surrounding it. The 27 dishes of the Very Large Array – a Y-shaped series of radio telescopes in New Mexico – use its power to look at explosive events on the Sun's surface. On the island of La Palma, the Swedish solar telescope looks at sunspots, providing the most detailed images of these blotches ever obtained and enabling solar astronomers to see and photograph details as small as 70 km wide on this ball of gas 1.49 million km across.

For the Soho (Solar and Heliospheric Observatory) satellite stationed in a gravitationally stable point on the sunward side of the Earth, there is no such thing as sunrise. Its sensors monitor various events and track sound waves rippling through the Sun's outer layers that tell us much about what lies beneath its 6000 degree surface. Scattered across the solar system many space probes sample the particles that constantly stream off it.

In a sense, this is the golden age of solar science. We have made remarkable discoveries, we have amazing new instruments and satellites that can probe the Sun's innermost secrets, and we have learned its fate, and possibly our own as well.

A scientist we shall come to know wrote in the early twentieth century that it was not too much to hope that in the not too distant future we shall be competent to understand so simple a thing as a star. Today we think that we do. He also wrote that we humans were bits of stellar matter that got cold by accident, bits of a star gone wrong – as we shall see.

Other suns

There are an almost uncountable number of other stars like our Sun throughout the Universe. It is an average star, a Goldilocks star if you like, neither too hot nor too cold. It is long-living and well behaved, which is perhaps the reason we are here. And it is on its own, whereas most stars are not.

Look across the vastness of space to our nearest stellar neighbours – to the Alpha Centauri system. This is 4.35 light years away ('only' being about 41 million million km) and it consists of three stars orbiting each other. Alpha Centauri A and B form a binary star, taking 80 years to circle one another. Alpha Centauri C (often called Proxima Centauri as it is the closest to us) is a dwarf star much further out and may even leave the system completely in a million years or so. Alpha Centauri A is very similar to our Sun, designated by astronomers as a type G2 yellow dwarf. It is the fourth brightest star in the sky and is so bright only because it is close by.

There are other so-called solar analogues – stars almost identical to our own Sun: LQ Hydra, HD 44594, HD 190406, 16 Cyg A and the most similar, 18 Scorpii, to name a few. The story of the Sun is therefore not unique, it has been and is being repeated across the length and breadth of the cosmos. In the starfields of our own galaxy and in the smudges of light from the most distant galaxies ever seen there will be stars like it. Barely anything that happens in this universe is not witnessed by a solar-type star. The Sun's life is a typical one, but no less remarkable for that.

Until relatively recently we were not sure if there were planets circling other stars, but in the past decade or so our knowledge has developed. We now know that very many of the closest stars have large, gas-giant worlds like Jupiter in orbit around them, and we surmise that there may be smaller, rocky worlds like Earth as well. In the future space-based observatories may be able to detect them. We can also see disks of dust around young stars that we believe will form planets. The Sun's retinue of planets is certainly not unique. We do not know for sure, but I believe that the Universe is teeming with life, much of which will live under the light of a yellow dwarf star like our Sun. Stars like it may in fact be the life-givers of the cosmos, responsible for creating the conditions for life to originate and flourish. If ever we do contact alien intelligences, intellects born under a foreign star, we may find that we have many things in common – perhaps one of them would be our parent Sun.

We can see stars like our Sun being born and we can see them die and from these observations we know the fate due to the Sun. Any extraterrestrial intelligence we may find would be old, for the chances are greater of finding long-lived civilisations than youthful ones like ourselves that have only recently acquired the technology to search the skies for signs of life or signal our presence to the cosmos. Perhaps these lifeforms will tell us of their lives around their star, and how one day they had to leave it because stars like our Sun that were born when the Universe was young have now come to the end of their useful lives.

Nevertheless, our Sun is nowhere near the end of its life. To tell the story of its birth we first have to survey the Universe and put our little star into its cosmic context.

The oldest recorded solar eclipse? 3340 BC.

CHAPTER THREE

The silken threads of space

B e they against the warm velvet of a summer's night or the cold and dark of November night skies, on a clear night you can see a few thousand stars with the unaided eye. They occupy only a small, nearby part of our Milky Way galaxy; telescopes reveal a much vaster realm of light from billions of galaxies and each galaxy represents the combined light from billions of stars.

Astronomers know there is much more out there than that which shines. Most of the universe is dark, made of matter that gives off no light and about which we know frustratingly little. We know it is there because of its gravity, but we cannot see it. The view from planet Earth shows us that we live in a Universe where stars and planets are comparative rarities, made of the minor constituents or impurities of the cosmos. There is the mysterious dark energy (about 70% of the universe) and the undiscovered dark matter (25%). Atoms make up only about 5% of matter. There is more to the Universe than meets the eye.

Despite this puzzling yet attractive situation (how dull it would be if we knew almost everything about the makeup of the cosmos), I regard stars as being as fundamental as atoms. When we gaze out across the depths of space, be it to the suburbs of our own galaxy or across the vast reaches of intergalactic space, we see the light from stars. It is through collections of stars – galaxies – and the groupings that galaxies make that we are able to map the Universe. And it is from stars that we get planets and life itself.

Although stars are almost everywhere, it was not always so. To begin the story of the Sun we have to go back to a time when the Universe was featureless and dark – the time before stars.

In the beginning

We now realise that all space, time and energy began with the Big Bang, estimated to have occurred around 13.7 billion years ago. But the youthful universe was not at all like the starry nights we know today. At the Big Bang all was hot beyond any form of measurement. Energy and matter, space and time, were indistinguishable and interchangeable. Within the first second the superhot gas, referred to as a plasma – a substance we will discuss often because the Sun is made of it – may have cooled enough for particles called quarks that had condensed out of the energy to combine and form protons and neutrons – the building blocks of atoms. After about three minutes, a small portion of the neutrons bonded with protons. For a few hundred thousand years the universe remained extremely hot at around a billion degrees, too hot for ordinary atoms to form.

Three to four hundred thousand years passed in this way, with the hot plasma eventually cooling and allowing atomic nuclei to hold onto electrons to make the atoms hydrogen and helium. It was then that the Universe entered the dark ages in which all that existed were gas, dark matter and, crucially, gravity.

Today, radio telescopes can detect the echo of the Big Bang in the form of an omnidirectional glow called the cosmic microwave background radiation. This echo originated about 380,000 years after the Big Bang when the first atoms formed and its properties tell us that the early Universe was at that time remarkably smooth. Tiny density fluctuations in the expanding and cooling plasma gas may have led to uneven concentrations in the primordial distribution of matter in the Universe, of which around nine-tenths was comprised of dark matter.

Today, dark matter is separate from ordinary matter (the stuff you and I are made of), having its home chiefly, we believe, in the outer regions of galaxies and in the spaces between the galaxies, but in those early times the two were intertwined. Gravity caused dark matter 'seeds' to form into a network of filaments and sheets. In a way, during these early phases of its expansion, the Universe spun a web with strands of diffuse matter acting like silken threads. So as the web was spun by dark matter's gravity, hydrogen and helium formed into forests of gas, which then fragmented into vast clouds of about 100,000 to one million solar masses that may have measured around 30 to 100 light years across and contained much dark matter.

As the gas clouds contracted under their own gravity, the compression would have heated the gas to temperatures above 1000 degrees. Hydrogen atoms would have paired up within the hot gas to create molecular hydrogen, which would then help to cool the densest parts of the gas cloud by getting rid of energy in the form of infrared radiation. Ordinary matter could then have collapsed into flattened, rotating clumps – possibly shaped like disks that resembled miniature spiral galaxies. In this way, ordinary matter was segregated from dark matter, which does not emit radiation and therefore easily loose energy and so remaining scattered throughout space. Inside the disks of ordinary matter the densest gas clumps continued to contract and eventually some underwent a runaway collapse to form the first stars. But the first stars were unlike those we see today. Our Sun is descended from giants.

The first stars

The first star-forming gas clouds were much warmer than the gas clouds that form stars today. They lacked the dust grains and heavier elements – so common today – that work to cool such clouds. The result was that the first stars to shine in the Universe were exceptionally massive and bright, and they changed the course of cosmic history. The significance of these first stars, appearing between 100 million and 250 million years after the Big Bang, was profound: they rebalanced the ecology of the cosmos.

The first stars were larger and hotter than those we see today but they obtained energy in basically the same way as the Sun does. As we shall see, this is from nuclear reactions whereby lighter particles combine to form heavier ones, liberating a little bit of energy when they fuse. Computer simulations suggest that the first stars had surface temperatures of about 100,000 degrees – about 17 times hotter than the Sun's surface. This means that the first starlight in the universe would have been intense, mainly ultraviolet radiation. It would have begun to heat and remove the electrons – a process called ionisation – from the hydrogen and helium gas around the first clusters of stars. The radiation would have created a growing bubble of ionised gas that grew as more and more stars formed. Eventually the bubbles of ionised gas would merge, allowing the power of the ultraviolet radiation to be felt almost everywhere.

The first stars would have had relatively short lifetimes – only a few million years – before they exploded. The explosion of a star – a supernova – is one of the Universe's most dramatic events. Today they are so bright that we can see them far across the cosmos; how much more cataclysmic must they have been when the first, giant stars tore themselves asunder? They were so bright and energetic that they would have run through their nuclear fuel rapidly. As they consumed this fuel, heavy elements, mostly iron, accumulated at their cores. Iron is important because it cannot burn, it marks an endpoint. Thus there comes a time when the energy from the core decreases and the outward pressure of radiation can no longer hold the weight of the star above it, which begins to fall. After a lifetime of millions of years the star's life is all over in just a few seconds. During the collapse matter is forced together under incredible pressure and electrons and protons fuse into neutrons. A shower of ghostly particles called neutrinos are produced in this process and they flood out into space at the speed of light, leaving behind the expanding entrails of a star so bright that, for a while, one star going supernova can have the brightness of ten billion suns.

The heavy elements made inside the star by nuclear burning are now scattered throughout space. They have a marked effect on the formation of the next generation of stars because they are much more effective than hydrogen in cooling clouds and allowing them to collapse into stars. This more efficient cooling allowed the formation of stars with smaller masses and may also have boosted the overall rate at which stars were born. It is possible that the pace of star formation did not accelerate until after heavy elements had been scattered like cosmic fertiliser. It was therefore the second generation of stars that were primarily responsible for lighting up the Universe in a cosmic renaissance.

Perhaps some of those early stars are still around. Some may be hiding in obscure corners of the cosmos, but some may be hiding in plain sight. The discovery a few years ago of an extremely ancient star designated HE 0107−5240 in the Milky Way's outer reaches demonstrated that stars that are less massive than the Sun can form from gas that lacks very many heavy elements. This was unexpected, as most current theoretical calculations indicate that it should have been very difficult to form low-mass stars shortly after the Big Bang. The existence of HE 0107−5240 suggests that there must be other ways of achieving the necessary cooling. Remarkably, some of the witnesses to those early times may still be around, shining faintly below easy detection as dwarf stars in far reaches of our galaxy's stellar halo.

So the stage was set for the birth of the Sun. We will consider how this may have come about, but first we have other stories to tell.

The four-pointed disk of Shamash has been a symbol of the Sun throughout the world for millenia.

CHAPTER FOUR

The shape of truth

Thousands of people in Japan, the 'land of the rising sun', go to the Futamigaura seashore at Ise to await the dawn on the first day of the year. The silence is interrupted only by the breaking of waves. As the pre-dawn light comes there can be seen a pair of Mateo-iwa rocks in the sea, one larger than the other. The large rock is said to represent a male and the smaller one a female, while a rope connecting them is the eternal bond that unites the divine couple. This is the Gateway of Heavenly Deities – the Shinmai-Dorii. On the large rock is the universal sun door – the symbol of immortality.

As the Sun rises the onlookers join hands, clap twice and bow in order to draw the attention of the sun goddess. They close their eyes and pray until the Sun has risen high enough in the sky that it is no longer red. They clap once more and leave. But they do not go far, they walk to the shrine of Amaterasu on the seashore. It is said that it was from the celestial cave that the sun goddess emerged to illuminate the world with the sunlight of spiritual salvation. In the small wooded shrine the sun icon is a tiny round mirror, which together with jade jewellery and a bronze dagger form the three sacred objects of Japanese mythology.

In England we do things differently. Festivities at Stonehenge in Wiltshire go on throughout the night. A disparate collection of people gather at the stones for the summer solstice: New Age revellers, Druids (although they have about as much to do with the stones as the New Agers), Wiccan worshippers, onlookers, police. They are all waiting for that one famous moment.

Stonehenge was built between 2800 and 1500 BC and for most of the time since then it has been a ruin. What must it have been like in its glory days? For much of the time between then and now the sunrise over the stones must have gone unwitnessed, such was the disinterest of centuries. But today the car park is full with all manner of vehicles and a cacophony fills the air. As the pink glow of dawn begins to show there is a scramble for a place near a large stone known as the Heel Stone that marks the point of sunrise on 21 June, the summer solstice.

The Heel Stone stands to one side of the axis of the henge. The origin of the name of this stone is uncertain. The name 'Heel' has also been recorded as 'Hele', giving rise to suggestions that the word may be derived from the Greek 'helios' or 'sun.' In Stone Age times the Sun would not have risen over it, as is popularly supposed, but to the left of it as viewed from the centre of the sarsen circle. In 1979 a stone hole to the left of the Heel Stone was uncovered, and this is now catalogued as stone 97. This may have been a former setting of the Heel Stone itself, but it could be more likely that the midsummer Sun rose between the Heel Stone and its now lost companion. The two uprights would have framed the first gleam of the Sun precisely, and the rising disk would then have grazed the tip of the Heel Stone.

The tip of the Sun appears above the horizon. The Druids wave their mistletoe and chant. The tension rises, the police stir. But come back to this spot in six months' time and you will be on your own to witness what is perhaps a more important event. The days are shorter and the air colder, the landscape is now harsh. Winter was a painful struggle for prehistoric man. Some of them must have witnessed the Winter Solstice and the midwinter sunset, framed between the uprights for the Great Trilithon. This is the shortest day (21 December) and after it the days will grow longer and the warmth will eventually return. Stonehenge is a midwinter marker saying that you have survived the worst, the Sun will soon return and bring with it the spring.

One thing that intrigues me about Stonehenge is something that I know is lost and gone forever. The stones have been carefully placed to map out the movement of the Sun and the Moon as they rise and set at various points on the horizon in cycles that take many years. Yet the people who built Stonehenge had, as far as we know, no written language. So they must have passed the knowledge of how to place the stones down the generations by using stories, poems and songs. I wonder what they were like, those ancient songs of the Sun, moon and stones? We will never know.

Throughout the world stones and temples are oriented at the Sun. Its influence permeated all societies. Since the dawn of history all civilisations have paid primaeval homage to it.

For the ancient Hindus the sun god was one of the principal triad of deities, called Surya in their Vedas or Books of Divine Knowledge. 'All that exists was born from Surya, the God of gods,' states the Rig Veda of ancient India. 'He who dwells in man and who dwells in the sun is one and the same,' says the Taittiriya Upanishad.

Polished convex discs of bronze with solar symbols on their nonreflecting side have been found in Siberia, having been used widely among the peoples of the Eurasian steppe. There is a stone tablet 3000 years old depicting the sun god Shamash in his temple with the solar disc before him and with a Babylonian king being led into his presence.

In the tomb at Shang-ts'un-ling in Honan, China, which is eighth century BC, there are believed to be the earliest mirrors in China. Dragons are interwoven with zoomorphs whose tails turn into volutes – symbols of the Sun.

We have always known that the Sun was part of our beginnings. Australian aborigines believe that during *wongar* or dreamtime there was a cave in which slept a beautiful woman – the Sun. The 'Origin People' of the Incas in Peru were thought to have come out of a cave, as their deity, Virachoca, created them from painted stone dolls. The sun god Utu, in Sumerian myth, helped Dumuzi, the young husband of the deity Inanna, to escape from the netherworld by turning him into a stone. Bhavisya Purana states that it was out of the orb of the Sun that Vishvakarma, the divine sculptor of the Hindus, chiselled the human form.

Petroglyphs from Kazakhstan, carved in the Bronze Age and now found on shattered rocks lying in the beds of mountain streams, show rays emanating from a circle; Mohenjo-daro seals found in the Indus Valley show animals emerging from the Sun's disc as it was thought mankind had done. A fourth-century BC orator, Aeschines, tells of a custom in ancient Greece in which, on their wedding eve, girls went on a sunny morning to bathe naked in the Skamandros river.

More than any other object in the heavens the Sun is intertwined with our lives and culture. The gesture of reassurance when the Buddha shows the palm of his hand is also a sun symbol that he called 'the fiery heart, my kith and kin'. To the ancient Greeks the shape of truth was the shape of the Sun, but the ancient peoples of the Indus Valley thought the Sun was square and conceived of the swastika sun symbol.

To the east of Lake Titicaca on the Collao High Plateau of Bolivia stands something that would not look out of place at Stonehenge. It is called the Gateway of the Sun – the door to immortality. In the Church of Santa Maria in Rome is a monolith called Bocca della Verita, a massive marble sun disk with a grim human face. People who tell lies are advised not to insert their hand into its mouth or the 'avenging justice of the Sun' will bite it off. Alexander the Great aspired to rule the world 'like the Sun' and Pharaoh Akhenaton sought to unite his people under the all-embracing light of the solar disc, the Aten. The Aztecs called themselves people of the Sun. The Sun has been the centre of the lives of all humans. They survived because they knew its yearly rhythm.

Sun worship

The worship of the Sun, although not peculiar to any one time or place, received its greatest prominence in ancient Egypt. There the daily birth, journey and death of the Sun were dominating features of life. One of the most important gods of Egyptian religion was Ra, the sun god, who was considered the first king of Egypt. The pharaoh, said to be the son of Ra, was the sun god's representative on earth.

In Mesopotamia, where sun worship was also very important, the sun god Shamash was a major deity and was equated with justice. In Greece there was Helios. The Greeks believed, just like the Egyptians believed of Ra, that Helios drove the Sun across the sky from east to west in his golden chariot every day. After sunset the Sun sailed back across the ocean. The influence of the Sun in religious belief also appears in Zoroastrianism, Mithraism, Roman religion, Hinduism, Buddhism, and among the Aztecs of Mexico, the Incas of Peru and many Native Americans.

Malina was and continues to be the sun goddess of the Inuit people, who live in Greenland. Malina and her brother, the moon god Anningan, lived together. They got into a terrible fight and Malina spread dirty, black grease all over her brother's face. In fear, she ran as far as she could into the sky and became the Sun. Annigan chased after her and became the Moon. This eternal chase makes the Sun alternate in the sky with the Moon.

Tonatiuh was a sun god for the Aztecs. They believed that four suns had been created in four previous ages and one of them had died at the end of each cosmic era. Tonatiuh was the fifth sun and the present era was his. He was in charge of the Aztec heaven, called Tollan. Only dead warriors and women who died in childbirth could be received in Tollan. Tonatiuh was responsible for supporting the Universe. To prevent the end of the world, Aztecs believed it was essential to maintain the strength of the sun god by offering him human sacrifices. The victims were usually prisoners captured in the frequent wars that Aztecs fought against their neighbours. The sacrifices were intended to secure rain, harvests and success in war. The most common form of sacrifice practised by Aztecs was to tear out the heart of a living body and offer it to the Sun.

Inti was considered the sun god and the ancestor of the Incas. In the remains of the Inca city of Machu Picchu, it is possible to see a shadow clock that describes the daily course of the Sun personified by Inti.

Chinese people believed that there existed ten suns that appeared in turn in the sky during the Chinese ten-day week. Each day the ten suns would travel with their mother, the goddess Xi He, to the Valley of the Light in the east. There, Xi He would wash her children in the lake and put them in the branches of an enormous mulberry tree called fu-sang. From the tree, only one sun would move off into the sky for a journey of one day, to reach the mount Yen-Tzu in the far west.

Tired of this routine, the ten suns decided to appear all together. The combined heat made life on Earth unbearable. To prevent the destruction of the Earth, the emperor Yao asked Di Jun, the father of the ten suns, to persuade his children to appear one at a time. They would not listen to him, so Di Jun sent an archer, Yi, armed with a magic bow and ten arrows to frighten the disobedient suns. However, Yi shot nine suns, leaving only the sun that we see today. Di Jun was so angry about the death of nine of his children that he condemned Yi to live as an ordinary mortal on Earth.

More than 3000 years ago Shamash was a sun god in Mesopotamia, between the valleys of the Tigris and Euphrates rivers. Since he could see everything on Earth, he also represented the god of justice. He and his wife, Aya, had two important children. Kittu represented justice and Misharu was law. Every morning the gates in the east open up and Shamash appears. He travels across the sky and enters the gate in the west. He travels through the underworld at night in order to begin in the east the next day. In Babylon, located in the south of Mesopotamia, the symbol of Shamash was a solar disk with a four-pointed star inside it.

Mesopotamia offers us the ancient epic of Gilgamesh, probably composed around 2000 BC. In this ancient Sumerian story Gilgamesh, king of Uruk, sets out on a quest for immortality to the Garden of the Sun, the land of everlasting life. To reach it,

Gilgamesh must pass through the Sun's gate in the mountain of the horizon. The setting Sun disappears there and emerges from it at sunrise. A pair of terrifying scorpion-people stationed at the gate of heaven guard the Sun's path.

The ancient Persians bowed to Mithra as the Sun, for it was said, 'May he come to us for protection, for joy, for mercy, for healing, for victory, for hallowing. Mithra will I honour with offerings, will I draw near to us as a friend with prayer.'

Indians to this day revere the Sun. Its Vedic names grew into some sort of active personality. We can follow step by step in the Vedic hymn the development that changes the Sun from a mere luminary into a creator, preserver and ruler. As the Sun sees everything and knows everything, he is asked to forgive and forget what he alone has seen and knows. He may be Indra, Varuna, Savritri or Dyaus, the shining one.

Some measure of the Sun's importance can be seen in the case of Amenhotep IV, an Egyptian pharaoh of the 18th dynasty who ruled in the fourteenth century BC. Husband to the beautiful Nefertiti, he was possibly the father of Tutankhamen, an otherwise obscure child ruler whose tomb somehow lay undiscovered until the twentieth century. Up until his time, Egypt paid homage to many gods: Osiris, Isis and Set among them as well as Ra, already mentioned. But Amenhotep had a wish to combine them into a single deity, the Sun, which he called Aten-Ra. He renamed himself Akhenaten, the servant of Aten-Ra. Akhenaten may not have acted purely out of religious feelings. One effect of the new religion was to dispossess the old priests, and Akhenaten was able to consolidate his power.

Nevertheless, his changes were short-lived. Akhenaten died in the city named after him, presumably of natural causes. The city of Akhetaten was abandoned and left in ruins, his brief experiment apparently ended as abruptly as it began. Akhenaten was despised by his successors, who all but erased him from history. On its rediscovery and examination in the nineteenth century, his tomb lay empty save for some debris and the smashed remnants of the heretic pharaoh's red granite sarcophagus.

Out of myth and legend came science and understanding. The ancients worshipped the Sun and noticed its behaviour and their records provide us with the first useful pieces of information about it – the first clues to its real nature.

Eclipse by Lucy Whitehouse

CHAPTER FIVE

The month of Mercedonius

*I*t was the Sun that taught us how to keep time and regulate our lives. Its daily passage across the sky changes only a little from day to day, but eventually we notice the changes as the seasons progress. Ancient man knew that. During the summer the Sun was at its highest in the sky, but there came a point when it had reached its highest and it started to traverse the sky a little lower each day, until it reached its lowest, at midwinter.

Although the cycle of the seasons was obvious, devising a calendar was not. There are only three regular measurements that can be used as a basis for a calendar and they do not coincide. Two of these are the day and the year, but given that there is not an even number of days in a year, how do you, with crude if any measuring devices, measure the number of days in a year?

The first calendars used the third cycle in the sky, which could be relied on. It was easier to count the 29 days between full moons and use that as a basis for timekeeping. There are some very ancient lunar calendars: 29 notches carved on a 28,000-year-old bone found in Africa and 29 dots found in a curious pattern in the cave paintings of Lascaux in the Dordogne in France, painted some 20,000 years ago when Europe was still in the grip of ice. However, lunar calendars have disadvantages. There is not a whole number of lunar months in a year so you cannot divide up the year, with its changing seasons, using the Moon. What was required was a calendrical system in which the dates stayed in synchronicity with the seasons and did not occur in one season one year and another some time later. The rainy season was always at the same time, so the date of its onset should be as well. The same applied to the annual flooding of the Nile River.

Egypt was the first ancient civilisation to realise the error of the Moon and embrace the Sun, making the change over 6000 years ago. How they did it we are not entirely sure. They estimated that the solar year was close to 365 days, which led to a calendar of 12 months of 30 days each, with an additional five days that Egyptian mythology says were added to the year by the god Thoth. These became the holidays of Osiris, Isis, Horus, Nephtys and Set.

The Egyptians had calculated that the solar year was actually closer to 365.25 days, but instead of having a single leap day every four years to account for the fractional day (the way we do now), they let the one-quarter day accumulate. After 1460 solar years, or four periods of 365 years, 1461 Egyptian years had passed. This meant that as the years passed the Egyptian months fell out of sync with the seasons, so the summer months eventually fell during winter.

When Rome emerged as a world power the difficulties of making a calendar were well known. The Romans believed that even numbers were unlucky so their months were 29 or 31 days long, with the exception of February, which had 28 days. But four months of 31 days, seven months of 29 days and one month of 28 days added up to only 355 days. Therefore the Romans invented an extra month called Mercedonius of 22 or 23 days, which was added every second year. Even with the month of Mercedonius it was a clumsy system. The Roman calendar eventually became so far off reality that Julius Caesar, advised by the astronomer Sosigenes, ordered a reform in 45 BC. One year was made 445 days long by imperial decree to bring the calendar back in step with the seasons. Then the solar year (with the value of 365 days and 6 hours) was made the basis of the calendar. The months were 30 or 31 days in length, and to take care of the 6 hours every fourth year was made a 366-day year, a leap year.

This was a considerable improvement, but it still had its problems. The Julian calendar is still 11.5 minutes longer than the actual solar year – the time it takes for the earth to go around the Sun – and after a number of centuries, the 11.5 minutes added up. By the fifteenth century the Julian calendar had drifted behind the solar calendar by about a week, so that the vernal equinox was falling around 12 March instead of around 20 March. Pope Sixtus IV decided that another reform was needed and called the German astronomer Regiomontanus to Rome to advise him. Regiomontanus arrived in 1475 but died shortly afterwards, and the pope's plans for reform died as well.

In 1545, the Council of Trent authorized Pope Paul III to reform the calendar. Most of the mathematical and astronomical work was done by Christopher Clavius. The immediate correction was that Thursday 4 October 1582 was to be the last day of the Julian calendar. The next day would be Friday 15 October. For long-range accuracy, Vatican librarian Aloysius Giglio's idea was adopted: every fourth year is a leap year unless it is a century year like 1700 or 1800. Century years can be leap years only when they are divisible by 400 (e.g. 1600 and 2000). This rule eliminates three leap years in four centuries, making the calendar sufficiently accurate. In spite of the revised leap year rule, an average calendar year is still about 26 seconds longer than the Earth's orbital period. But this discrepancy will need 3323 years to build up to a single day.

Most Catholic countries quickly changed to the Pope's new calendar in 1582, but Europe's Protestant princes chose to ignore the papal bull and continued with the Julian calendar. It was not until 1700 that the Protestant rulers of Germany and the Netherlands changed to the new calendar. In Great Britain the shift did not take place until 1752, and in Russia a revolution was needed to introduce the Gregorian calendar in 1918. In Turkey, the Islamic calendar was used until 1926.

Shamash – The Mesopotamian Sun God saw everything on Earth.

CHAPTER SIX

The Hill of the Witch

Loughcrew, just inside the Meath border at Oldcastle in the Irish midlands, is a lonely, old place with a feeling of ancient peoples who, if you do not look up at the hills, may live there still but who quickly vanish behind the stones when you try to observe them. Stretching over four hilltops, it is also known as Sliabh na Caillighe, meaning 'hill of the witch'. Scattered around are the remains of over 30 passage tombs dating back to around 3300 BC in the Neolithic or new Stone Age, the time of Ireland's first farmers.

Two tombs on different hills stand out. The climb to both sites is very steep and you need stout footwear and care. Inside them are highly decorated stones with swirls and spirals similar to those found further south at the slightly younger sites of Newgrange and Knowth. One tomb has inside it a cruciform chamber, a corbelled roof and some of the most beautiful examples of Neolithic art in Ireland. During the vernal and autumnal equinox, people gather at dawn in the tomb to watch the Sun's rays enter the chamber and illuminate the interior. The sunlight is shaped by its passage between the stones and descends the backstone illuminating carved solar symbols.

Here, in November 3340 BC, the Sun seemed to set twice as the locals witnessed an eclipse. It was subsequently recorded on three carved stones. Looking at them you can see the eclipse. On one there is a large spiral with a smaller one emerging out of the side, just like the first bite of the Sun's disc by the encroaching moon.

In modern times when the Sun rises on 8 November, its rays strike a stone pillar inside the cairn. But due to the passage of time and the slow wobble of the Earth's axis, the alignment in 3340 BC was slightly different. The next morning after the eclipse, 1 December 3340 BC, the Sun would have beamed directly onto the pillar, indicating that it was placed in its location on that morning to commemorate the Sun's return.

Further south, at dawn on the winter solstice, just after 9 am, the Sun begins to rise across the Boyne Valley from Newgrange over a hill known locally as Red Mountain. At four and a half minutes past nine, the light from the rising Sun strikes the front of Newgrange and enters into the passage through the roof box, which was specially designed to capture the rays. The roof box was discovered in 1963 but its solar significance was not recognised until 1967. For the following 14 minutes, the beam of light stretches into the passage and eventually reaches the central chamber, where in Neolithic times it illuminated the rear stone – the sunstone – of the central recess. Then the shaft of light disappears and once again the chamber returns to darkness.

Loughcrew and Newgrange are not the only places in the British Isles where Stone Age man captured the rays of the Sun. A mysterious prehistoric tomb on the island of Orkney has a special 'light box' cut into its roof that allowed a shaft of light to herald the start and end of winter. The Crantit tomb was discovered in 1998 when a tractor disturbed a series of flat stones just below the surface. These turned out to be the roof slabs of an underground tomb, hidden below a hillside for over 5000 years. A team from Glasgow University's archaeological field unit soon realised the importance of the find.

The small tomb consists of three chambers in a clover-leaf formation and is almost invisible from above ground. It is highly unusual, as most chambered cairns in Orkney and elsewhere were built on the surface as prominent mounds. But during the excavation one of the archaeologists, a skilled stonemason, noticed something unusual about the slabs of stone that formed the roof. A notch had been cut in the roof to allow a ray of sunlight to penetrate the tomb, but only at certain times of the year. In October and again in February, at the beginning and the end of what was no doubt a harsh northern winter, the Sun would have thrown a shaft of light along the length of the tomb. Strange carvings were found on the upright stone pillar that supports the roof. If you look closely you can see geometric patterns and symbols carved into the rock. It is thought they may even have been coloured with primitive paints and pigments. No human remains were found in the tomb's central chamber, but bones from four individuals were found in side chambers: a mature woman, a girl, a child and one too degraded to identify. Fragments of pottery were also found but no ornaments. Neolithic tombs were frequently used over and over again, but not this tomb. The entrance was heavily blocked up, both inside and out, suggesting that no reuse was intended. Perhaps it tells us something about a cultural change in the treatment of death and burial 5000 years ago. It may be significant that the tomb took the human remains back to the Earth but also let in the rays of the Sun.

Why did those people, struggling to survive among the forests and hills of northwest Europe not long after the retreat of the ice, build such things? They were a great investment of time and energy, commodities that could have been put into finding food or improving shelters. Clearly, their connection with the Sun was important and was something beyond their day-to-day cares, so they carved and heaved stone, watched and wondered and they left us symbols whose meanings we can only guess. But these creative people were not alone in observing the Sun at the dawn of civilisation.

A reproach from heaven

The civilisation of China is the world's oldest. The traditional blessing of the emperor, Wansui, which means 'ten thousand years', underestimates its antiquity. Aspects of the emerging civilisation – rice cultivation, pottery, carving – go back more than 10,000 years. Around 4000 BC farming communities developed along the Huang He River that travels for 3000 miles from the Himalayas to the sea. Yellow soil is washed down with it that give its more familiar name, the Yellow River.

Civilisation in China is said to have begun with the founding of the Hsia (which means summer) dynasty. We know very little about this period and have not much more than a list of the 17 kings of the dynasty. According to some Chinese historians, it was the last of the sage kings, Yu, who legend says ruled with perfect wisdom, clarity and virtue, who founded the dynasty. It lasted for about 400 years until it was overthrown by T'ang in 1766 BC, when he, predictably, started a new dynasty, the Shang.

At the start of the Hsia dynasty Chinese astrologers already had sophisticated observatory buildings in which to sit all night to observe the heavens. Ancient Chinese astronomy was primarily a government activity. The role of the astronomers or astrologers (for there was no distinction) was to keep track of the motions of the Sun, Moon and planets and explain what they foretold for the divine emperor.

There are a pair of oracle bones from the Shang dynasty that have written on them, 'On day *kuei-yu* the sun was eclipse in the evening; is it good? On day kuei it was enquired: the sun was eclipse in the evening; is it bad?'

The fact is, as the oracle bones intimated, that the Sun often disappeared to punish humanity for its wrong doings or poor advice given to the emperor. Throughout the ancient orient eclipses were thought to be a direct result of a leader's actions and naturally the reproach from heaven was regarded as important. The book *History from the former Han Dynasty* relates the warning of one Chinese ruler: 'If the sun at its rising is like a crescent and wears a crown like the moon: the king will capture his enemy's land; evil will leave the land, and (the land) will experience good ...' Another report says, 'On the day of *chi-ch'on*, the sun was eclipsed and it became dark in the daytime. The Empress Dowager was upset by it and her heart was ill at ease. Turning to those around her she said, "This is on my account." '

The emperor's divine mandate was written in the stars for all to see. One Chinese commentator wrote of an eclipse, 'When the moon covers a large portion of the sun the consequences will be serious, misfortune will fall upon the head of state – the next year the emperor died.' When the prince of men is not virtuous, a reproach appears from heaven and earth: ' ... Our experience in governing has been short, so that we must have not been correct in our acts; hence during *wu-shen* there was an eclipse of the sun and an earthquake. We are greatly dismayed!'

So it was that one day, without warning, a dragon began eating the Sun, a critique of current events if ever there was one. Courtiers gossiped, advisers panicked and army chiefs commanded their men to beat their shields and fire arrows into the air to scare the dragon away, peasants banged on tree trunks and shouted, children ran home from the rice fields. Fortunately, the army's swift action worked as within minutes the dragon took heed and left the Sun alone. Nevertheless, the emperor was not pleased: the court astrologers should have predicted this. A search was begun to find them.

The culprits, Hi and Ho, were found the worse for drink. Because they had put the world in danger and almost allowed the Sun to be swallowed up by a dragon, they were put to death. It may be true, it may be a myth. It was later written:

Here lie the bodies of Ho and Hi,
Whose fate, though sad, is risible;
Being slain because they could not spy
Th' eclipse which was invisible.

Portents of doom

It is clear that the life of a court astrologer in China or Japan was a respected if not a secure one. In early Japan, fewer than 20% of solar eclipses predicted to occur before AD 1600 were actually observed. This was because if a forecast eclipse did not take place, its absence was credited to the emperor's high virtue; as a result his royal power was enhanced and the court astrologer rewarded. However, if a solar eclipse occurred that had not been foreseen, there was no time to ward off danger and the astrologer might very well face death. So it helped to lengthen the life of the poor astrologer if he predicted as many eclipses as possible and hoped for the best. Even so, Hi and Ho may not have been the only unfortunates.

Lest today we be comforted by our scientific enlightenment of how the Universe works, recall that in 1914 Russian peasants fled to their churches in horror and despair after a total solar eclipse occurred in the Ukraine. And at the 1922 Australian eclipse, aborigines believed that the astronomers who had gathered at a nearby site to view the event were trying to capture the Sun in a net. In 1994 Indians high in the Andean Mountains lit fires to keep the Earth warm during a solar eclipse.

Almost every civilisation noted and recorded eclipses. In 1928 a small group of French archaeologists journeyed with seven camels, one donkey and some porters towards a small hill known as Ras Shamra in Syria. After a week excavating the site they discovered a cemetery. In the graves they found Egyptian and Phoenician artwork and alabaster as well as some Mycenean and Cypriot materials dating back to the second millennium BC. This was as they expected since the region, being on the shore of the Mediterranean, was a crossroads of culture and influence.

This site, Ugarit, had a very long history. A city was built on the site in the Neolithic period around 6000 BC. The oldest written evidence of this city is found in some texts from the nearby city of Ebla and were written around 1800 BC. Both Ebla and Ugarit were under Egyptian hegemony, but in the period 1200–1180 the city declined and then mysteriously came to an end.

The texts discovered at Ugarit were written in four languages: Sumerian, Akkadian, Hurritic and Ugaritic. Many were found in the royal palace and the house of the high priest, and some in the private houses of leading citizens. But the most remarkable discovery was a collection of tablets carved with an unknown cuneiform script. The style of writing discovered at Ugarit is known as alphabetic cuneiform, a unique blending of an alphabetic script (like Hebrew) and cuneiform (like Akkadian). Most likely it came into being as cuneiform was passing from the scene and alphabetic scripts were becoming more established. Ugarit's influence is still with us, especially for those who study the Bible, for the voices from this city have echoes in the Old Testament. Several of the Psalms were adapted from Ugaritic sources and the story of the flood has a near mirror image in Ugaritic literature.

A clay tablet found at Ugarit has been identified as a description of a total solar eclipse that occurred on 5 March 1223 BC. Another tablet reads: 'The day of the moon of Hiyaru was put to shame: the sun went in, with her gatekeeper, Rashap.' It is clear that by the eighth century BC, the Babylonians and other cultures were keeping a systematic record of solar eclipses, and may even have been able to predict them based on numerological rules.

One date of an eclipse referred to in the Bible is known for certain: ' "And on that day," says the Lord God, "I will make the Sun go down at noon, and darken the Earth in broad daylight" ' (Amos 8 : 9). 'That day' was 15 June 763 BC. The date of this eclipse is confirmed by an Assyrian historical record known as the Eponym Canon. In Assyria each year was named after a different ruling official and the year's events were recorded under that name in the Canon. Under the year corresponding to 763 BC, a scribe at Nineveh recorded this eclipse and emphasised the importance of the event by drawing a line across the tablet. These ancient records have allowed historians to use eclipse data to improve the chronology of early biblical times.

Eclipses as portents of doom persisted down the ages. According to an early entry in the Anglo-Saxon Chronicles:

In this year King Henry went across the seas at Lammas and the next day, when he was lying asleep on board ship, the day grew dark over all lands, and the Sun became as if it were a three-nights-old Moon, with stars about it at midday. People were very much astonished and terrified and said that something important would be bound to come after this – and so it did, for during that same year the King died. ...

However, there is more to this than is first apparent. Historians, astrologers and monks were often more influenced by the needs and wants of the royal family or the state than the requirement of accuracy. King Henry died in AD 1135, but the solar eclipse that was credited with causing his demise actually occurred two years earlier.

The spectacle of an eclipse was impressive to ancient poets. A fragment of a lost poem by Archilochus goes: 'Nothing there is beyond hope, nothing that can be sworn impossible, nothing wonderful, since Zeus, father of the Olympians, made night from midday, hiding the light of the shining Sun and sore fear came upon men.' This has been identified as a description of the total solar eclipse of 6 April 648 BC.

But rather than an eclipse, it was the fall of a meteorite in 467 BC, said to be the size of a millstone, in daylight at Thrace near Aegospotami, and the Arigos Potamoi (Arigos river) that was a crucial event in the early understanding of the Sun. The stone was shown off to visitors by the people of Abydos on the Hellespont. The philosopher Anaxagoras was supposed to have predicted its fall, but if true this must have been coincidence. No doubt he went to see that meteorite near Arigos Potamoi and no doubt he was deeply impressed with it. It convinced him that heavenly bodies were not divine beings and he believed that the Sun was made of a mass of red-hot iron no larger than the Peleponnessus, roughly 160 km. Because of these views Anaxagoras become the earliest figure in history to be indicted for atheism and condemned to death in absentia. 'Nature,' he stoically remarked about his accusers, 'long ago condemned both them and me.'

Anaxagoras believed that the Sun was rotating at such a high speed that it was extremely hot. A meteorite was a broken piece of the Sun, he maintained. He is an important figure in the early understanding of the Sun because he also developed the first scientific theory of the solar eclipse. Although Thales had predicted the solar eclipse, he did not know how to explain it. Anaxagoras's theory of the solar eclipse, that it is caused by the Moon coming between the Earth and the Sun, was correct. He also correctly believed that the Moon did not have its own source of light, but reflected the light of the Sun.

Anaxagoras lived in Athens for about 30 years, a time spanning the golden age of classical Greek culture. He had a rational mind and a calm and detached perception of the world. Near the end of his life he left Athens, perhaps because he was again charged with impiety, and retired around 433 BC to the Hellespontine city of Lampsacus. There he was welcomed, surrounded by students and honoured by the citizens. When he died in about 428 BC he was given a public funeral, and the citizenry inscribed a tribute on his tomb:

Here lies Anaxagoras, whose picture
of the order of the universe
came closest to the truth.

Shortly before his death his followers asked him what he would consider an appropriate way to honour his memory. He replied that students should be given the month in which he died as a vacation each year. Anaxagoras was so highly respected in Lampsacus that his death was observed as he wished for well over a century. Euripedes wrote in 431 BC:

The moon is eclipsed through the interposition of the earth, and sometimes also of the bodies below the moon. The sun is eclipsed at the new moon, when the moon is interposed ... Anaxagoras was the first to set out distinctly the facts about eclipses and illuminations.

It was at this time that someone used the Sun to measure the size of the Earth. It happened at a place that today we call Aswan, but the Ancient Egyptians called it Swen and considered it to be not only the end of the world, but also the sacred source of the Nile. For centuries it was the gateway to Africa and the lands of Nubia. The Copts called it Souan, meaning 'trade', from which the present-day Aswan is derived.

In the time of the Ptolemies and ancient Egyptians, the town of Syene stood southwest of the present-day Aswan. Syene was situated on the island of Yeb, or today's Elephantine Island, and was a major trade centre with Nubia and the rest of Africa. Its quarries, with those of Rohannou, were the principal ones of Egypt; they supplied a certain kind of red granite called syenite, out of which were cut the obelisks, monolithic temples, the colossus and so on. From the time of the ancient empire royal Egyptian envoys went there to look for the stone destined for the sarcophagus of the pharoh. These quarries were in full activity during the Roman epoch and syenite was exported throughout the empire.

Another celebrated place in Syene was a pit, which was incorrectly thought to have been placed exactly under the equator. It was chosen by Erastostenes as the starting point for his measurement of the surface of the Earth in 230 BC. Erastostenes noticed that on midsummer's day a vertical stick cast no shadow at Syene but cast a seven degree shadow at Alexandria. If the earth were flat, then the sticks would cast the same shadow. If the earth were round, they would not. Knowing the angle and the distance between Syene and Alexandria he calculated, with remarkable accuracy, the circumference of the Earth. Erastostenes became the chief librarian at Alexandria but went blind at 80 and the following year chose to starve himself to death.

Halos of light

Eclipses of the Sun have left their mark, sometime in ways we have forgotten. Religious images often show Jesus Christ and the saints with haloes. The use of the halo, or nimbus, originated with the pagan Greeks and Romans to represent their sun god, Helios (or Apollo). Later artists adopted it for use in Christian images. It has its origin with the Sun shining behind one's head, or possibly with the corona or halo of light seen around the Sun during an eclipse.

The solar corona may have been recognised as early as the time of Plutarch who remarked on the unusual light surrounding the Moon's disc during an eclipse in AD 83. The corona is the hot, extended outer atmosphere of the Sun. It is far too faint to be seen against the blinding brightness of the solar disk itself, but becomes visible at times of total solar eclipses when the solar disk is obscured by the Moon. The first recorded observation of the corona during a solar eclipse was in Ancient China on oracle bones: 'Three flames ate up the Sun and a great star was visible.'

In one account written by Plutarch, a certain Lucius says that even if the Moon during an eclipse does cover the Sun entirely, the eclipse does not have duration or extension, but a kind of light is visible about the rim that keeps the shadow from being profound and absolute.

While the solar corona is visible at any solar eclipse, the first explicit mention of it was made by the Byzantine historian Leo Diaconus (ca. 950–94), as he observed the total eclipse of 22 December 968 from Constantinople. His observation is preserved in the Annales Sangallenses and reads:

> ... at the fourth hour of the day ... darkness covered the earth and all the brightest stars shone forth. And it was possible to see the disk of the Sun, dull and unlit and a dim and feeble glow like a narrow band shining in a circle around the edge of the disk.

The growth of astronomy

It was in Ancient Greece that the Sun ceased being a god. Anaximenes in the sixth century BC thought that the Sun was a flat body in the sky supported by air. Xenophanes around the same time, however, thought that the Sun and stars were fiery clouds, illuminated by their own motion through the cosmos; the Sun was renewed every morning and the stars each evening. But for many Greeks the sun god Helios rose each morning from the lake of the Sun, a gulf in the huge river Oceanos that surrounded the earth. From the 5th century BC the Greeks associated the Sun with the god Phoebus Apollo.

But, Claudius Ptolemy was having none of that. He made astronomical observations from Alexandria in Egypt during the years AD 127–41. In a way he was the most influential astronomer of all time and he was not only an astronomer. He was a mathematician and geographer who codified the Greek Earth-centred view of the universe.

The Moon revolves around the Earth, so it must be that the Sun and the planets do the same, argued Ptolemy. The heavens are the realm of perfection, unlike the corruption on Earth. The most perfect form of motion is a circle, so Ptolemy had the planets orbit the Sun and the planets orbit the Earth in circles and in circles moving in circles, in a scheme of epicycles that is at the same time beautiful and artificial. It was a theory that lasted for 1400 years, perhaps a great deal longer than any modern scientific theory can be expected to survive without radical change.

To the Greeks and to Ptolemy the Sun was important, but the Earth and mankind were more important so it was clear that they were the centre of the Universe. God had placed his children at the centre of all things and all things were created in tribute to God. Ptolemy believed that the planets and the Sun orbited the Earth in the order Mercury, Venus, Sun, Mars, Jupiter, Saturn – a system that became known as the Ptolemaic system. He described his work in a book called the *Mathematical Syntaxis* (widely called the *Almagest*). A 13-volume epic, it contains a myriad of information ranging from Earth conceptions to Sun, Moon, and star movement as well as eclipses and a breakdown on the length of months. The *Almagest* also included a star catalogue containing 48 constellations, using the names we still use today.

In the first book of the *Almagest*, Ptolemy describes his geocentric system and gives various arguments to prove that, in its position at the centre of the Universe, the Earth must be immovable. He showed that if the Earth moved, as some earlier philosophers had suggested, then certain phenomena should be observed in consequence. In particular, Ptolemy argued that since all bodies fall to the centre of the Universe, the Earth must be fixed there at the centre, otherwise falling objects would not be seen to drop towards the centre of the Earth. As a result of such arguments, the geocentric system became dogmatically asserted in western Christendom until the fifteenth century, when it was supplanted by the heliocentric (Sun-centred) system of Nicolaus Copernicus. Ptolemy's *Almagest* shares with Euclid's *Elements* the glory of being the scientific text longest in use.

To be fair, the planetary theory that Ptolemy developed is a masterpiece. He created a sophisticated mathematical model to fit observational data, which before his time was scarce, and the model he produced, although complicated, represented the motions of the planets fairly well.

After the fading of the ancient Greeks, the next steps in science were taken by a different society. Muslim armies, which had been unified by Mohammed's doctrine, began their conquests in 636. They seized Syria, Iraq, Mesopotamia and Egypt, taking Alexandria, home of Ptolemy and what remained of the great library. As Europe lapsed into the Dark Ages, Muslim influence extended through Turkey, North Africa, Spain and as far east as the borders of China and India. The Arabs absorbed ideas from mathematics, astronomy and other sciences from the cultures and regions they conquered. By the year 750 the wars had subsided and a time of relative peace prevailed. Scholars from different regions gathered in Baghdad and caliph Al Mamun established a House of Wisdom in the city. Many Greek texts were translated into Arabic for the first time. Most of the Arabs' work was in the field of mathematics and astronomy, but there were also significant advances made in physics.

Historical literature is often careless with its description of scientific contributions from the Arab world in the Middle Ages. Many see this period as merely a time of the preservation of Greek knowledge until the scientific revolution dawned in Europe. But the Arabic contribution was important to the rise of science. Calculation was a delight to Moorish scholars. Their critiques and improvements of Greek ideas acted to refine previous concepts. If the Greeks were the authors then the Arabs were the editors, and in addition they invented the numbers we now use, including the zero. It was in Arabic studies where the modern scientific method first appeared. It also appears that many of the scientific ideas in Europe after the Middle Ages were really borrowed from or influenced by Arab accomplishments.

NICOLAI CO
PERNICI TORINENSIS
DE REVOLVTIONIBVS ORBI=
um coelestium, Libri VI.

Habes in hoc opere iam recens nato, & ædito,
studiose lector, Motus stellarum, tam fixarum,
quàm erraticarum, cum ex ueteribus, tum etiam
ex recentibus obseruationibus restitutos: & no=
uis insuper ac admirabilibus hypothesibus or=
natos. Habes etiam Tabulas expeditissimas, ex
quibus eosdem ad quoduis tempus quàm facilli
me calculare poteris. Igitur eme, lege, fruere.

ἀγεωμέτρητος οὐδεὶς εἰσίτω.

Collegij Brunsbergensis Societatis Iesu

Norimbergæ apud Ioh. Petreium,
Anno M. D. XLIII.

*Reverendo Dno Georgio
doctori canonico Varmiensi
amico suo ——
K. Losius dd.*

Seen on his deathbed – Copernicus's 'Revolutions'.

CHAPTER SEVEN

The Dethronement of God

It was the Sun that dethroned God and destroyed mankind's special place in the universe, causing a fundamental shift in the way people viewed themselves. The position and status of the Sun became the key point, scientific and philosophical, in an argument between the established wisdom of Ptolemy and the received wisdom of the scriptures on the one hand, and the discoveries and speculations of the rising power of science on the other. Until the medieval period, with the exception of a few enlightened but ultimately uninfluential Greeks, the prevailing cosmology was that of Ptolemy and Aristotle. Their works did not just describe the Universe, they came to define it. The Earth was the centre of the universe and the Sun, along with all the planets, went around it.

This is the world view that was destroyed in the sixteenth and seventeenth centuries. The first blow was dealt by Nicolaus Copernicus (1473–1543), who published his landmark book *De Revolutionibus Orbium Coelestium* in 1543. This book brought about one of the two fundamental shifts in man's status. *The Revolutions*, as it is called, stands alongside Darwin's *Origin of Species* as one of the most important scientific books ever written.

In the spring of 1543 Copernicus lay dying of a brain haemorrhage in Northern Poland when friends brought him a package from the German printer Johannes Petreius in Nuremberg. It was a book: *On the Revolutions of the Heavenly Spheres*. He had been reluctant to publish his theory. It is said that he was afraid of persecution by the Church, but that is wrong. It seems he was chiefly afraid of being mocked. He had the support of Pope Paul III and the first presentation of his system was given (in 1532) by the Pope's private secretary, in the Vatican gardens.

By the sixteenth century, astronomers realised that Ptolemy's Earth-centred system had great problems. It could not easily account for the fact that Mercury and Venus are never seen to stray too far away from the Sun, while Mars, Jupiter and Saturn sometimes appear to move backwards in their orbits. Copernicus read the ancient Greek writers and rediscovered their Sun-centred theories. He realised that these would lead to a massive simplification of Ptolemy's model, and he set out to put their views on a firm mathematical footing.

He was born on 19 February 1473, in Thorn (now Torun), Poland, to a family of merchants and municipal officials. He went to Jagiellonian University in 1491, studied the liberal arts for four years without receiving a degree, and then, like many Poles of his social class, went to Italy to study medicine and law. Before he left, his uncle had him appointed a church administrator in Frauenberg (now Frombork); this was a post with financial responsibilities but no priestly duties. In January 1497 he began to study canon law at the University of Bologna while living in the home of a mathematics professor, Domenico Maria de Novara, who it was first inspired the somewhat dreamy Copernicus to take an interest in geography and astronomy. In 1500 he gained permission to study medicine at Padua, the university where Galileo was to teach nearly a century later. It was not unusual at the time to study a subject at one university and then to receive a degree from another, less expensive institution. So Copernicus, without completing his medical studies, received a doctorate in canon law from Ferrara in 1503 and then returned to Poland to take up his administrative duties.

Back in Poland he published his first book, a Latin translation of letters on morals by a seventh-century Byzantine writer, Theophylactus of Simocatta. Sometime between 1507 and 1515, he completed a short essay known as 'The Commentariolus', which was not published until the nineteenth century. In it he laid down the principles that Domenico Maria de Novara had taught him about a Sun-centred planetary system. He distributed it among his friends asking them write down any comments; this way he began to realise what people would make of the theory and if the time was ripe to proclaim it further.

After moving to Frauenburg in 1512, Copernicus took part in the Fifth Lateran Council's commission on calendar reform in 1515, and wrote a treatise on money in 1517. It was not until 1530 that he seriously began working on *De Revolutionibus Orbium Coelestium*, which was finished by a pupil of his who was eager to see the work in print.

Copernicus himself originally gave credit to Aristarchus of Samos when he wrote, 'Philolaus believed in the mobility of the earth, and some even say that Aristarchus of Samos was of that opinion.' Interestingly, this passage was crossed out shortly before publication, maybe because Copernicus decided that his treatise should stand on its own merit. Nevertheless, if it stood, how could a transient body like the Earth, filled with meteorological phenomena, pestilence and wars, be part of a perfect and inscrutable heaven? If it stood and were literally true, then what was biblical authority worth if the Earth was not the centre of all things?

Andreas Osiander (1498–1552) was a theologian experienced in shepherding mathematical books through production. He urged Copernicus to present his ideas as purely hypothetical, and he introduced certain changes to the book without permission. He added an unsigned 'letter to the reader' directly after the title page, which said that the hypotheses contained within were not actually true because astronomy was incapable of finding the causes of heavenly phenomena. At this Copernicus's pupil Rheticus's rage was so great that he crossed out the letter with a great red *X* in the copies sent to him. But Osiander's act was kept quiet and was not made public until Johannes Kepler revealed it in his book the *Astronomia Nova*, in 1609. In addition, the title of

Copernicus's work was changed from the manuscript's *On the Revolutions of the Orbs of the World* to *Six Books Concerning the Revolutions of the Heavenly Orbs* – a change that mitigated the book's great claim.

Did Copernicus know of the letter to the reader? He may have held the book in his crippled hands before he died, but what else he thought we will never know.

De Revolutionibus was freely circulated for 70 years. It took the observational skills of the next generation of astronomers Tycho Brahe and the mathematical abilities of Johann Kepler to refine the new system, replace circles with ellipses, and put the heliocentric model on a proper mathematical footing. Then came Galileo and the telescope which provided observational evidence that the Sun was the centre of the Universe, not the Earth. On 5 March 1616, the work was forbidden by the Congregation of the Index 'until corrected' and in 1620 these corrections were indicated. Nine sentences, by which the heliocentric system was represented as certain, had to be either omitted or changed.

Copernicus presented a new planetary model with the Sun placed at the centre and all the planets (including the Earth) orbiting the Sun. Under this arrangement the orbital speed of planets decreases steadily outwards as they get further away from the Sun, and the outer sphere of fixed stars is motionless. He introduced his heliocentric model in order to do away with the complications of Ptolemy's model, with all its cycles and epicycles introduced to explain the apparent motion of the Sun and planets across the sky. The Copernican system maintained a clear distinction between the Sun and the 'fixed' stars, distributed on the fixed, outermost sphere of the Copernican cosmos. This was rejected by the next generation of Copernicans following Kepler and Galileo. René Descartes (1596–1650) who, in his 1644 book *Principia Philosophiae*, asserted that the Sun was but a star that had formed at the centre of a primaeval vortex.

The Sun was now just a star, like the ones that can be seen on any starry night.

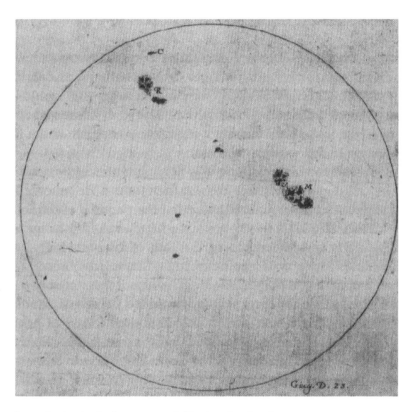

Imperfection in the heavens – Galileo's solar sketches.

CHAPTER EIGHT

A hole in the sky

An Inuit creation myth says that when the world was young and all was dark, the raven escorted man into the sky. They came to a round hole in the sky that was glowing like fire. This, the raven said, was a star.

Once astronomers thought there were holes in the sky, regions of inky blackness in an otherwise lush field of stars. We now know them as dark clouds of molecules. Inside is a relatively high concentration of dust and molecular gas that absorbs practically all the visible light coming from background stars. If you use your imagination to enter one of them, you might indeed feel as if you were entering a hole in space. These starless and bible-black regions make the interiors of molecular clouds some of the coldest and most isolated places in the Universe.

One of the most notable of these dark nebulae – Greek for cloud – is found towards the constellation Ophiuchus. It is known as Barnard 68. No stars are visible in the centre indicating that it is relatively nearby, about 500 light years away and half a light year across. It is not known exactly how molecular clouds like Barnard 68 form, but it is a cosmic egg, a stellar nursery.

Stars develop in the cold dark clouds of gas and dust that are scattered throughout the cosmos. However, while much has been learned over the past two decades about the birth of stars there are still many mysteries about the process. Recently astronomers figured out a way to peer inside a dark cloud by looking at the light from more distant stars. By measuring the amount of starlight absorbed by various regions, they calculated its density and the temperature differences from the core to its outskirts. The results revealed an object on the verge of collapse.

Barnard 68 is a good candidate in which to study star birth, as it has a particularly simple shape and well-defined edges. The cloud contains about twice as much mass as our Sun but is widely distributed: Barnard 68 covers a linear distance about 12,500 times bigger than the distance from the Earth to the Sun. The cloud, therefore, is many billions of times less dense than the air you are breathing.

Even so, the stars behind Barnard 68 are mostly obscured, their light scattered by its dust. It's an effect similar to that which makes a sunset on Earth appear red. Blue light, having a short wavelength, is the most scattered and doesn't get through, while longer-wavelength red light passes through more easily. In fact, Barnard 68 is dense enough that if you could reshape it and slip it between us and the Sun, the Earth would be plunged into total darkness.

With a large telescope, like the behemoths at the La Silla and Paranal observatories in Chile, operated by the European Southern Observatory, detailed photographs have been taken of Barnard 68. Long exposures have revealed 3708 background stars. By measuring the colours of these stars, after their light had been scattered by dust, it is possible to probe the dust cloud and tell where the gas and dust are and what gas it is; gas and dust are known to stick together. Barnard 68 does seem to be poised at the beginning of a collapse, expected to take place in about 100,000 years. As it contracts it will eventually reach a new balance and a new star – one very similar to our Sun at its birth several billion years ago – will grace the galaxy.

That stars come from clouds of gas and dust was suspected quite early on. By the closing decade of the eighteenth century, the increasingly powerful reflecting telescopes built by the German-born English astronomer William Herschel (1738–1822), had revealed the existence of a number of diffuse cloud-like structures, dubbed nebulae. Inspired by these observations, the French astronomer and mathematician Pierre Simon de

Laplace (1749–1827) put forth his nebular hypothesis, according to which the Sun and solar system formed from the gravitational collapse of an initially slowly rotating, large but diffuse gas cloud. Laplace's cosmological ideas were described in a popular work, published in 1796 and entitled *Exposition du système du monde*. It marked a turning point in the history of science, since therein he categorically rejected the biblical account of the creation of the Universe, and offered instead a physically based theory that, in its main thrust if not in all details, is valid to this day.

However, not all gas and dust clouds will form stars. These arise only when a cloud of gas is cold enough for gravity to overwhelm it and cause it to collapse. Gas cannot cool if it is exposed to the strong flux of ultraviolet light that bathes the galaxy, so star-forming clouds are those that can shield themselves from this radiation, mainly because they are full of dust. This stardust is composed of small silicate grains produced by previous generations of giant stars, and it plays a vital role in forming planets. In essence, the Earth is a compressed ball of stardust.

While the dust is essential for the birth of stars, it obscures what is going on, making observations of star birth difficult. While it efficiently prevents ultraviolet radiation from getting to star-forming clouds – and allowing them to cool to just ten degrees above absolute zero, the temperature required to initiate star formation – it also prevents the radiation from newly formed stars from getting out. Star formation has not so much been shrouded in mystery as shrouded in dust, and much of the progress in recent years has been because of new ways to peer into the dust.

Gravity will try to pull the gas and dust together. But the material also exerts an outward pressure. The warmer the material, the more active it is and the more it pushes outward. In some nebulae, where the temperature is cool enough and the force of gravity overcomes the outward pressure, the gas and dust become more tightly packed and eventually a gravitational collapse occurs. Clumps of material form called protostars. Soon the temperature at the centre of the protostar has risen enough to start thermonuclear fusion, the process which converts hydrogen to helium that powers the stars, including our Sun. The heat from this burning creates thermal pressure, which keeps the protostar from collapsing further under its own weight. Once this question of balance has been addressed, the star can shine for perhaps billions of years.

Looking deep into the Orion nebula – the most famous stellar nursery, a spectacular gas cloud easily visible to the unaided eye in the winter constellation of Orion the hunter – indicates to some astronomers that stars are born in much more crowded environments than had been thought. In the Orion nebula many stars are just a few light weeks' distance from each other.

The birth of the Sun

Astronomers think that our Sun was formed in a similar way from a giant cloud that was rotating slowly, about 5000 million years ago. Perhaps in the Sun's case its birth was induced, as the collapse could have been triggered by another stellar death in the form of a shock wave from the explosion of a supernova. Or possibly a cluster of hot blue supergiant stars, termed OB associations, produced an interstellar wind from their tremendous outpourings of radiation, which compressed the surrounding material to form stars.

The Sun went through a tempestuous youth, whipping up strong winds from its surface that swept the solar system clear of whatever gas had not been incorporated into a planet. But then the Sun settled down. From studying rocks, fossils and Antarctic ice, scientists think that it has been brightening over time by about 30% to its present luminosity, but has declined dramatically in its production of intense ionizing ultraviolet and extreme ultraviolet radiation. The Sun's rotation has also slowed and its wind has subsided to the contemporary solar wind, of which more later.

By studying the radioactive decay of certain elements in rocks, geologists can determine how old those rocks are. The oldest that have been found on Earth are between 3000 and 4000 million years old, but the oldest meteorites that have been discovered are about 4500 million years old. Assuming that the Earth and Sun were formed at about the same time as the oldest meteorites and making allowance for some period of time for the formation of meteorites, gives the age of the Earth and roughly that of the Sun to be about 4700 million years.

Our Sun is therefore a third of the age of the Universe. It is a second-generation star, made of material that was once inside another star but was scattered into space after its demise. Everything that is not hydrogen or helium – the carbon in our bodies, the calcium in our bones, the oxygen we breathe – was made inside a star and drifted through space as part of a gas cloud for aeons until it was gathered up in the Sun and the disk of debris around it out of which the planets formed. We are stardust.

The building frenzy of the new solar system as planets formed erased most of the chemical memory of its origins, but some dust remains untouched, locked inside meteorites, frozen in time. Since their discovery a decade ago these precious grains, which weigh only a billionth of a gram, have fuelled an explosion of research into the evolution of our galaxy and the formation of our solar system. It seems they have come from carbon-rich red giant stars, typically one to five times the mass of the Sun, that were approaching the end of their lives. They were bloated with enormous but relatively cool atmospheres in which silicon and other grains formed. They also had strong winds that drove the gas and dust grains into space. The grains tell us something about the gas and dust cloud out of which the Sun formed. Puzzlingly, it seems that it may have been slightly unusual or highly evolved stars that migrated outwards from the centre of the galaxy before spilling their contents over our own particular solar birth cloud.

Stars very seldom form alone. Usually many dozens of stars are formed from a single big cloud of hydrogen gas. The Pleiades, or the Seven Sisters seen on winter evenings, are one such group of stars. These sister stars eventually move off in different directions, and after about 1000 million years it is no longer possible to tell that these stars ever belonged together.

One of the many intriguing observations made by the Hubble Space Telescope is of a young double star designated HK Tau. It was the first example seen of a young double star system with an edge-on disk around one member of the pair. The thin, dark disk is illuminated by the light of its hidden central star. It is a puzzle that stars appear to be twinned at birth much more frequently than at later times. I wonder, was the Sun like HK Tau? Did it once have a partner? Our Sun may have sister stars out there, but we don't know where they are and probably never will.

ISTORIA
E DIMOSTRAZIONI
INTORNO ALLE MACCHIE SOLARI
E LORO ACCIDENTI
COMPRESE IN TRE LETTERE SCRITTE
ALL'ILLVSTRISSIMO SIGNOR
MARCO VELSERI LINCEO
DVVMVIRO D'AVGVSTA
CONSIGLIERO DI SVA MAESTA CESAREA

DAL SIGNOR

GALILEO GALILEI LINCEO
Nobil Fiorentino, Filosofo, e Matematico Primario del Sereniss.
D. COSIMO II. GRAN DVCA DI TOSCANA.

IN ROMA, Appresso Giacomo Mascardi. MDCXIII.

CON LICENZA DE SVPERIORI.

Letters on sunspots by Galileo – Rome 1613.

CHAPTER NINE

The Stourbridge prism

*I*saac Newton now enters our story, as indeed he does most stories of the development of physical science. It was he who carried out some fundamental experiments that showed that light from the Sun was made of a rainbow of colours.

One day he bought a prism at the Stourbridge Fair, sold as a curio by a lens grinder. He had it sitting on his desk and noticed how when the Sun shone on it, he saw different colours coming out. Did the prism change the light, or did the sunlight have lots of colours that the prism put into different places, he wondered.

He had been thinking about prisms and rainbows. He knew, like many in his day, that the edge of a lens in a telescope produced a rainbow set of fringes around an object being looked at because every lens at its edge is a prism. But the explanation, going back to the Greeks, did not ring true to Newton. They said that white light travels through the glass and is darkened a little at the thin end becoming red and is darkened more by the thick end becoming blue.

To find the answer, Newton used his blinds to get a very thin sunbeam to strike the prism. In doing this he saw that the separation of light was even clearer.

It was at first a very pleasing divertisement, to view the vivid and intense colours produced thereby; but after a while applying myself to consider them more circumspectly, I became surprised to see them in an oblong form; which, according to the received laws of Refraction should have been circular.

And I saw ... that the light, tending to one end of the Image, did suffer a Refraction considerably greater than the light tending to the other. And so the true cause of the light of that image was detected to be no other, then that light consists of Rays differently refrangible, which, without any respect to a difference in their incidence, were, according to their degrees of refrangibility, transmitted towards divers parts of the wall.

There was red, then orange, then yellow, then green and then blue. He was now sure that the light from the Sun had all these colours in it and that what the prism was doing was bending them into slightly different directions. So it was that Newton showed that sunlight can be separated into separate chromatic components via refraction through a glass prism.

Nobody believed him. In fact he withdrew from the debate about the rainbow, as was his character, and it was not until 1704 when he published his *Optics* that he set out his unarguable theory in full.

Such was the breadth of Newton's talent that he also measured the mass of the Sun. The mass of the Sun and its distance from the Earth are two very fundamental quantities that were only determined with reasonable accuracy in the eighteenth century. The first quantitative estimate of the Sun's mass is due to Newton, who presented the calculation in his *Principia Mathematica*, making use of his newly formulated law of universal gravitation. The ratio of Sun to Earth mass can be determined in principle, without knowing the actual value of the universal gravitational constant – the strength of gravity – which featured in his calculations. This only required knowledge of the orbital period of the earth and the radii of the Earth and Sun. In his first attempt Newton used an incorrect value for the size of the Earth's orbit and consequently underestimated the Sun to Earth mass ratio by more than a factor of ten (he had a Sun to Earth mass ratio of 28,700 instead of 332,945). But in later editions of his *Principia* he put the matter right and brought his estimate to within a factor of two of the true value.

It is impossible to overestimate the influence of Newton. In *The Prelude* William Wordsworth paid him homage:

> *And from my pillow, looking forth by light*
> *Of moon or favouring stars, I could behold*
> *The antechapel where the statue stood*
> *Of Newton with his prism and silent face,*
> *The marble index of a mind for ever*
> *Voyaging through strange seas of Thought, alone.*

Solar observers

It was in 1672 that the distance to the Sun was finally measured to any degree of accuracy by the first English Astronomer Royal, John Flamsteed, and the director of the Paris Observatory, Giovanni Cassini. The basic method was to triangulate Mars as seen from two widely separated places on Earth, because according to Kepler's laws, knowing the relative distances of the planets you just needed one distance to set the absolute scale for the rest. Cassini estimated a solar distance of 22,000 Earth radii or 149 million km.

A century or so later, the discoverer of Uranus, William Herschel, was an active solar observer with a gift for making telescopes and, sometimes, an overactive imagination. He was having problems cutting down the glare of the Sun so that he could observe it. In 1800, he wrote,

> before the late transit of Mercury over the disk of the sun, I prepared my 7-foot telescope ... As I wished to keep the whole aperture of the mirror open, I soon cracked every one of the darkening slips of wedged glasses, which are generally used with achromatic telescopes: none of them could withstand the accumulated heat.

To reduce the brightness of the Sun he experimented with filters made of various fluids, publishing in 1801 a report describing how:

> I viewed the sun with a skeleton eye-piece, into the vacancy of which may be placed a moveable trough, shut up at the ends with well-polished plain glasses, so that the sun's rays may be made to pass through any liquid contained in the trough, before they come to the eye-glass. Through spirit of wine, I saw the sun very distinctly.

He later wrote, 'I viewed the sun through water. It keeps the heat off so well, that we may look for any length of time' and 'I viewed the sun through Port wine, and without smoke on the darkening glasses.'

In attempting to diminish the Sun's brilliance by viewing it through coloured glass plates Herschel noticed something peculiar: that the red glass reduced the intensity of visible light more than it reduced heat and green glass reduced heat more than light. Herschel deduced that the Sun radiated both heat and light:

> The instrument I wished to adapt for solar inspection was a Newtonian reflector, with 9 inches aperture; and my aim was, to use the whole of it open. I began with a red glass; and ... took two of them together ... the eye could not bear the irritation, from a sensation of heat ... I now took two green glasses; but found they did not intercept light enough. I therefore smoked one of them ... they still transmitted considerably more light than the red glasses, they remedied the former inconvenience of an irritation arising from heat ... nothing remained but to find such materials as would give us the colour ... of a pale green light, sufficiently tempered for the eye to bear its lustre.

It was a crucial observation to which we will return later. Herschel had discovered that the Sun's colours, its spectrum, extend beyond the visible. He notes that with 'combinations of differently coloured darkening glasses ... some of them, I felt a sensation of heat, though I had but little light; while others gave me much light, with scarce any sensation of heat.' Always thorough, Herschel included instructions for properly smoking glass. 'Smoke from sealing wax is bad, pitch is worse, wax candles are good, tallow candles are better, whale oil lamps are as good as any.'

His son, John, was also a keen solar observer who designed a new form of eyepiece to aid in solar observing – the Herschel wedge, first described in his 'Results of Astronomical Observations ... at the Cape of Good Hope ... 1834–8', published in 1847. It was a simple unsilvered glass surface, wedge shaped to prevent reflections from the rear surface from reaching the eyepiece and getting into the eye. Herschel did not intend its use as an independent accessory, as it is sometimes employed today. It was described as used in conjunction with an unsilvered glass primary mirror that was concave on the backside, to form a double concave 'lens' that would refract and diverge the sunlight to disperse its heat (the back of the tube must be open as well, for the sunlight to escape).

These days few astronomers use an eyepiece like the Herschel wedge. The best way to reduce the problem of concentrated light and heat in a telescope is to not let the light and heat in in the first place, so the best device is a filter of a special foil-like material fitted over the front of the telescope. Never, never look at the Sun with the unaided eye or through any form of optical aid. Blindness can result. Always project the image through the telescope or binoculars, as we shall see that astronomers learned to do.

Life on the Sun

But back to John's father William Herschel, who had some rather peculiar ideas about life on the Sun. Writing in 1794, he said it was possible that living beings inhabited the interior of the Sun.

Curiously, over the ages there had been no shortage of speculation that it might be inhabited. The Greek Lucian played with the idea in his *True History*, the first real science fiction story:

> After crossing the river, we found something wonderful in the grapevines. The part which came out of the ground, the trunk itself, was stout and well-grown, but the upper part was in each case a woman, entirely perfect from the waist up. Out of their finger-tips grew the branches and they were full of grapes. When we came up, they welcomed and greeted us. They even kissed us on the lips and everyone that was kissed at once became reeling drunk. They did not suffer us, however, to gather any of the fruit, but cried out in pain when it was plucked. Some of them actually wanted us to embrace them, and two of my comrades complied, but could not get away again. They were held fast by the genitals, which had grown in and struck root. Already branches had grown from their fingers, tendrils entwined them, and they were on the point of bearing fruit like the others any minute.
>
> 'The king of the inhabitants of the Sun, Phaethon,' said Endymion king of the Moon, 'has been at war with us for a long time now. Once upon a time I gathered together the poorest people in my kingdom and undertook to plant a colony on the Morning Star which was empty and uninhabited. Phaethon out of jealousy thwarted the colonisation, meeting us halfway at the head of his dragoons. At that time we were beaten, for we were not a match for them in strength and we retreated. Now, however, I desire to make war again and plant the colony.'

Thommaso Campanella (1568–1639) also built an entire utopian narrative in his *City of the Sun*.

In 1795 Herschel maintained that the Sun was essentially a large planet with a solid surface, surrounded by two layers of clouds. An opaque lower layer shielded the solar inhabitants from the heat and light of the glowing upper layer which he thought was similar in nature to the Earth's aurora though on a grander scale. He wrote,

> The Sun ... appears to be nothing else than a very eminent, large, and lucid planet ... Its similarity to the other globes of the solar system ... leads us to suppose that it is most probably ... inhabited ... by beings whose organs are adapted to the peculiar circumstances of that vast globe.

Among others who shared this belief in the Sun as a life-bearing planet were the scientists François Arago and David Brewster. In fact, such weight did Herschel's views carry that serious proponents of an inhabited Sun were still to be found in the 1860s. More recently, with the realisation that alien life might be completely different to the terrestrial variety, scientifically informed writers have revisited the possibility of organisms living inside or on the surface of stars. Among these are David Brin in his novel *Sundiver2* and the late Robert Forward with his novels about a technological civilisation on a neutron star, *Dragon's Egg*.

Nobody takes the idea of life on the Sun seriously these days; or rather, almost nobody. In the 1977 science fiction novel *If the Stars Are Gods*, Gregory Benford and Gordon Eklund propose, for a while, that the Sun might have an intelligence. Earlier, in 1951, an elderly German engineer made a public wager that scientists could not prove life did not live in the Sun. (Given our ignorance of what forms life could take that should have been a safe, if rather pointless bet.) He argued that because sunspots are dark they were holes leading to the Sun's interior, where life could live at the Sun's core! His name was G. Bueren (1881?–1954) and he lived in Osnabruck. Bueren, however, was adjudged to have lost the bet and was forced by a German court to make payment to the Astronomische Gesellschaft. After Bueren's death, his estate owed the Gesellschaft money, which was finally used in the form of a 'Bueren Fund' legally established for training young German students of astronomy.

Nevertheless, some have argued that perhaps Bueren was correct, merely a bit premature! As we shall see in our exploration of the Sun's future, in about five billion years or so Earth will enter a gaseous holocaust caused when the Sun swells to become a red giant star. Earth will survive intact, not evaporating inside the Sun, and may eventually be ejected from the Sun into interstellar space. It is not wholly inconceivable that heat-tolerant microbial life forms just may one day live inside the Sun's now coolish outer atmosphere. We could even genetically engineer them to live there, if we so wished.

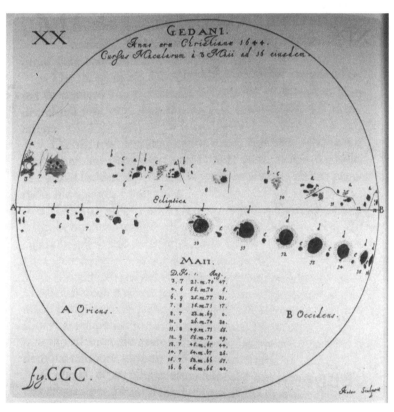

Scheiner showed how sunspots moved.

CHAPTER TEN

The fire at Mannheim

The key to understanding the Sun turned out to be in the flame, or rather the flames rising from an unfortunate fire. One evening in 1857 a fire was raging in the port city of Mannheim in Germany. Lying on the right bank of the Rhine, it is a city that had been destroyed many times as it swayed between the French, the Austrians and the Prussians and more than half of it was destined to be raised to the ground in World War II. Mannheim gave us the motor car thanks to a resident by the name of Carl Benz. But that night the flames of Mannheim opened the way to understanding what the Sun and the Universe were made of and it all came about thanks to a young man found scrambling around in the wreckage of a collapsed house, and the patronage of a wealthy prince.

Joseph von Fraunhofer was destined to have a profound impact on science. Born to a family of glassmakers, he was surrounded by optics from the time he was born. By the age of 19, Fraunhofer was making the best lenses yet produced.

He was born on 6 March 1787 in Straubing as the eleventh and last child of Franz Xaver Fraunhofer, a glazier. Seven of his brothers and sisters died early; his mother died aged 54 when Joseph was 10 and a year later his father died. He had to become an apprentice wood turner, but soon changed to a glazier since he was not strong enough for the work. He began in 1799 in the workshop of Weichselberger in Munich. Weichselberger forbade him to visit Sunday school and to read books, but fate intervened when, literally, the house fell down around him.

The collapse of two houses on 21 July 1801 changed Fraunhofer's life forever. Weichselberger's wife died in the rubble, but Joseph was found almost unhurt when he was rescued after four hours. On the scene of the disaster was Maximilian Joseph, later to be King Maximilian I, and Joseph von Utzschneider, a politician and entrepreneur. Just weeks earlier Utzschneider had left his political career to concentrate on his business: making fine instruments and optics. Taking a liking to Fraunhofer, Maximilian Joseph gave him some money and his protection. With it Fraunhofer bought a glass-cutting machine and a grinding machine and left Weichselberger's workshop. But unfortunately things did not work out and for a while he returned to his former master as journeyman. Nevertheless, the friends he had made around the ruins of the house continued to help him.

Joseph von Utzschneider and his friend Georg von Reichenbach had founded a workshop in Munich in 1802 to produce the geodetic instruments required to make maps. Reichenbach had studied at the military academy in Mannheim and had been sent to England to learn about James Watt's steam engines. During his time in England he also learned about optical instruments from Ramsden, Dollond and others. In the workshop he had some instruments on his shelves, almost ready to be sold except for the fact they had no lenses. Enter Fraunhofer, who developed new ways to make lenses better than any made before. He found the glass to be a major variable in a good lens, so he welcomed Utzschneider moving his workshop to Benediktbeuern, where he had founded a glass-melting workshop that should be able to produce glass of high quality. Fraunhofer learned the secrets of glass melting to add to his repertoire.

From then on there was no stopping him. He developed new types of glass and instigated a production plan that had one instrument per day leaving the factory. A list exists describing 37 instruments and includes prices offering a heliometer at 1430 Guilders, some 'comet seekers', astronomical telescopes, telescopes, loupes and prisms. A microscope with six objectives and two eyepieces is listed at 520 Guilders. Fraunhofer calculated, designed and tested every instrument, wrote the manuals and watched the disassembling and packaging process for the bigger telescopes. He was just 22 years old and his instruments were sold and put to use throughout Europe. The instruments he produced helped him make scientific breakthroughs of his own. He made a spectrometer that split the Sun's light into its component colours using a prism and he began exploring the solar spectrum, noticing that it contained mysterious dark lines. He usually focused his instruments on a candle in order to calibrate them, but one day he chose to focus on the Sun. He noticed that there were dark lines that were not present in the candle and went on to identify 574 different ones.

They had been seen before by British astronomer William Hyde Wollaston, but he did not pay them the same interest as Fraunhofer, who assigned them the letters A to Z, a nomenclature that is still used today. His drawing of the spectrum, which still exists, is wonderful to look at and the printing plate that he engraved is a masterpiece. He discovered that different elements were responsible for different lines along the spectrum and that the moon and planets gave the same pattern as the Sun, but curiously stars gave different patterns.

Fraunhofer continued his experiments and manufacturing until his death by tuberculosis (probably aggravated by his glassblowing) on 7 June 1826. His work on the solar spectrum had opened the way to determine not only what the Sun was made of but almost everything else in the universe as well.

In 1844 the French philosopher Auguste Comte – the founder of positivism, a philosophical system of thought maintaining that the goal of knowledge is simply to describe the phenomena experienced, not to question whether it exists or not – mused on what would be forever hidden and decided that mankind would never be able to deduce what the stars and the planets were made of. We could never visit them, so how could we know what they are made of? Auguste Comte was wrong and the fire at Mannheim helped prove him so.

The industry of optics

Thanks in part to Fraunhofer, optics in the early nineteenth century was a flourishing industry. Napoleon Bonaparte's passion for spyglasses had set surveyors and generals to order telescopes and the work of Herschel, who charted the southern skies from the Cape of Good Hope, inspired an interest in large telescopes. So a new breed of artisans emerged – opticians. One such was Jesse Ramsden of London. His eight-foot altitude-measuring circle built for the Dunsink Observatory in Dublin was a masterpiece, but it was delivered 23 years after its contracted deadline (when waiting for the manuscript of this book my publishers must have experienced a similar feeling to the Dublin astronomers). He was equally late for his social functions: he once arrived for a party at Buckingham Palace at the hour and day specified, but exactly one year too late!

A certain Robert Wilhelm Bunsen (1811–99) also played a crucial role in our understanding of the Sun. Bunsen devoted much of his career to chemistry. During his experiments he lost the sight of an eye and nearly died of arsenic poisoning. Apparently, one woman said of him, since he smelled terrible because of all the chemicals that he was working with, 'First I would like to wash Bunsen and then I would like to kiss him because he is such a charming man.' Bunsen never married and when asked, he would say he never could find the time.

The man after whom the Bunsen burner is named was born and educated in Göttingen, Germany, where he studied chemistry. Using a government grant, he toured Germany, France, Austria and Switzerland largely on foot, studying geology, visiting mines and factories, and meeting scientists.

With his fellow professor Gustav Robert Kirchhoff, who made so many fundamental discoveries about electricity, Bunsen was working on spectroscopy – the analysis of light – when one evening they saw from the window of their lab in Heidelberg the fire raging in Mannheim some 15 km to the west. Kirchhoff and Bunsen had developed their own spectroscope with which they began studying the spectral 'signature' of various chemical elements in gaseous form. Using their spectroscope they detected the lines of barium and strontium in the flames from Mannheim. This set Bunsen wondering whether they might be able to detect specific chemical elements in the spectrum of the Sun.

Kirchhoff extended Fraunhofer's work and was able to explain the dark lines in the Sun's spectrum as caused by absorption of particular wavelengths as the light passes through a gas. He found that when light passes through a gas, the gas absorbs those wavelengths that it would emit if heated. In 1859 he published an explanation of the Fraunhofer lines, in which he suggested that they are due to absorption of certain wavelengths by substances in the Sun's atmosphere. It was after all possible to tell what the Sun and stars were made of and a new era in astronomy had begun.

It all seemed to fit. As early as 1758 Andreas Sigismund Marggraf (1709–82) noted that soda colours flames yellow, and potash gives a lavender flame. William Henry Fox Talbot (1800–77), who pioneered photography and deciphered Assyrian inscriptions at Nineveh, distinguished red Lithium and Strontium flames using a prism. In 1854 David Alter of Freeport, Pennsylvania (1807–81), found that each element he studied had its own unique spectrum.

So on 15 November 1859 Bunsen wrote to a friend that he had found a means to determine the composition of the Sun and fixed stars with the same accuracy as we determine sulfuric acid, chlorine and so on, with our chemical reagents: 'Substances on the earth can be determined by this method just as easily as on the sun, so that, for example, I have been able to detect Lithium in 20 grams of sea water.'

Curiously, it was through spectroscopic observations of the Sun that the second most common element in the Universe was identified. There is a lot of helium on the earth but, because it is not a very reactive element, it is difficult to find if you do not know what you are looking for. The identification of helium was a remarkable achievement for nineteenth-century solar scientists.

A bright yellow line, close to the lines due to sodium but slightly shorter in wavelength, was observed in the Sun's corona during an eclipse in 1868. Although at first mistaken for the sodium lines, the slight difference in wavelengths was noted by two of the leading solar astronomers of the day, Frenchman P. J. C. Janssen and Englishman Norman Lockyer.

Tests carried out at the College of Chemistry in London by Lockyer were made to reproduce the lines, but it was impossible to find the element responsible for the strong yellow line. Thus in 1870 Lockyer suggested that it was due to a hypothetical element that he named 'helium', after the Greek sun god Helios. In 1882 it was observed in gases erupting from Vesuvius. Twenty-five years later, in 1895, William Ramsay confirmed the existence of helium when he managed to isolate it from the mineral cleveite. Lockyer was later knighted for this discovery.

So the secret of finding out what the Sun is made of lay in the solar spectrum. But let us leave for a while the development of instruments and techniques designed to find out what the Sun is made of and what its physical state is. The time has come to consider sunspots and their story, after a slight deviation concerning sunbathing.

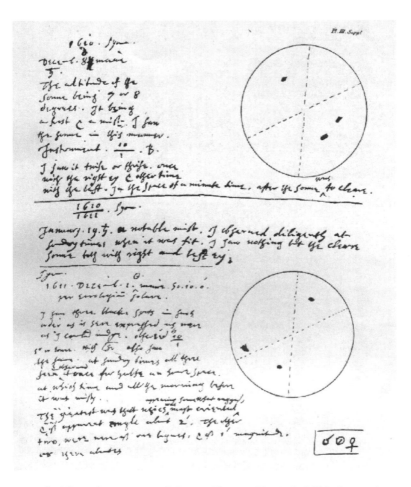

From the Thame's mists – Englishman Thomas Harriot's 1610 observations of the Sun.

CHAPTER ELEVEN

Sunlight robbery

The tribes of sub-Saharan Africa have a special relationship with the Sun. For many peoples the Sun is a source of light and warmth, bringing growth and fruition to the world. But in some areas of the world, the bountiful Sun also sterilises, desiccates and kills. In much of Africa the year is divided into two distinct seasons, one of rain and life, and the other ruled by the Sun's scorching power. For many African tribes the Sun is an avatar of excess, dispassionate about mankind. The Sun has both a kind and a cruel side.

To most people the Sun is a source of joy and refreshment, of enrichment and wellbeing. A sunny day is a good day when we seek the outdoors, some of the burden of indoor confinement is lifted and we are free, released from an indefinable restriction. But that is not always the case. There are times when the Sun is not our friend and times when its absence is keenly felt.

SAD

Throughout the centuries, poets and writers have described a feeling of sadness, melancholy and lethargy during the shortening days of autumn and winter, when the sunlight is pale and un-accompanied by warmth. It is perhaps no coincidence that many cultures and religions have winter festivals associated with candles or fire at about the time of midwinter, when the hope of spring grows in those who have suffered so far. Today, some call it 'the blues'.

For some people 'the blues' is not just a passing lowness that can be shaken off with a small application of willpower. They experience an exaggerated form of winter sadness, a depression and lack of energy that becomes debilitating. It is called seasonal affective disorder (SAD) and estimates are that it affects tens of millions each winter. Depression, lack of energy, increased need for sleep and weight gain are common symptoms. But it need not only occur in winter. Susceptible individuals who work in buildings without windows and exposure to the Sun may experience SAD-type symptoms at any time of year. Some people with SAD have mild or occasionally severe periods during the spring or summer. There is a smaller group of individuals who suffer from summer depression.

About 70–80% of those with SAD are women and the most common age of onset is in one's thirties, but cases of childhood SAD have been reported. The incidence of SAD increases with geographical latitude up to a point, but does not continue increasing all the way to the poles. There seems to be a relationship between a person's vulnerability and degree of light exposure. For instance, one person might feel fine all year in London, in southern England, but develop SAD when she moves to Aberdeen in the north of Scotland. Another individual may be symptomatic in Edinburgh, even though that is south of Aberdeen, but not in Plymouth in the far south of England. Some who work long hours inside office buildings with few windows may experience symptoms all year round. Some very sensitive people may note changes in mood during long stretches of cloudy weather.

It was in 1984 that a psychiatrist at the National Institute of Mental Health in the United States, Norman Rosenthal, published a report on the use of bright light therapy in patients with this disorder. Since then, other studies have confirmed his findings. Researchers are still investigating the way in which bright light can lift depression or reset a sleep cycle but one theory is that an area of the brain near the visual pathway, the suprachiasmatic nucleus, reacts to light by sending out a signal to inhibit the secretion of a hormone called melatonin. This natural hormone is made by the pea-sized pineal gland located just above the middle of the brain. During the day the pineal is inactive, but when the Sun goes down, or darkness occurs, the pineal is activated by the suprachiasmatic nucleus and it begins to produce melatonin.

When melatonin levels rise in the blood we feel less alert, sleep becomes inviting and melatonin levels stay elevated in the blood for about 12 hours. The amount of melatonin released at night varies among individuals, which is why 'night owls' are at their best in the evening, but it is somewhat related to age. Children on average secrete more melatonin than adults, which decreases further with age. However, research has shown that older people with sleep problems do not always have lower melatonin levels than people who experience normal sleep.

Some believe that there is a pathway from the retina to the suprachiasmatic nucleus. However, some recent and rather surprising research suggests that bright light applied to the back of an individual's knee could shift human day–night rhythms and the release of melatonin. This suggests that the bloodstream, not just the neurons of the visual pathways, might mediate the biological clock. Shining a bright light behind your knees at the right time may even help alleviate jet lag!

Artificial sunlight either in the morning or the evening can help with SAD. Initially researchers felt that full-spectrum light was required, but studies now suggest that regular fluorescent lights will work as well. Ultraviolet light can damage eyes and skin, so it must be filtered out. Many people still prefer full-spectrum (minus UV) light because it is the closest to natural lighting. One of the symptoms of SAD can be difficulty in waking up in the morning and so some find it helpful to have the artificial sunlight turn on just before they are supposed to wake up – a dawn simulator. Some use a bright light that is programmed to increase its intensity gradually such that it reaches its full intensity a set period before the individual is scheduled to awaken. Outdoor light, even when the sky is overcast, is best. There has been a study showing improvement in SAD symptoms when individuals take a one-hour daily walk outside.

Most totally blind people should owe no biological allegiance to the Sun except that they have to fit into the life rhythms of those who do. They have circadian (day–night) rhythms that are 'free running' (i.e. they are not synchronised to environmental time cues and oscillate on a cycle slightly longer than 24 hours). This condition causes recurrent insomnia and daytime sleepiness when the rhythms drift out of phase with the normal 24-hour cycle. It has been found that administration of melatonin can bring their rhythms under control. You would have thought that over millions of years of evolution, over millions of years of only sunlight and firelight as sources of illumination, we would have become exactly tuned to the day–night cycle? But curiously, no!

Humans and animals generally have innate sleep–wake cycles close to but not exactly 24 hours. They depend on the daily light–dark cycle to keep their circadian rhythms constantly reset to a regular 24 hours. However, if a human is left in a room with no light–dark cues, he or she will gradually shift into a sleep–wake cycle that is not exactly 24 hours long. What is more, the autonomous cycle length varies at different periods in the life span. Adolescents often have an innate cycle longer than 24 hours so that they have the desire to stay up late and sleep in when it is time to get up. The innate cycle then shifts closer to 24 hours for adults, but for the elderly, the autonomous sleep–wake cycle may be shorter than 24 hours, resulting in evening tiredness, sleep difficulty and waking too early.

Today we live in a 24-hour society. More than any other time in history, we sleep less and work and play more. Since Edison invented the light bulb, we have toyed with our biological clocks that tell us to sleep when it is dark. It is not surprising that this has begun to take its toll on us. Nearly all of us have had trouble sleeping at one time or another, whether we suffer from SAD or not.

Rickets

Samuel Pepys, the famous diarist and observer of London life, expressed his admiration for the company of Daniel Whistler in a number of his entries. On 13 January 1664, Pepys remarked that he met Daniel Whistler at a coffeehouse and had 'extraordinary good discourse' with him. Daniel Whistler was a remarkable physician. He had graduated from Leiden University in 1645. His thesis was on the disease called rickets and was the first printed book on the subject.

Rickets had been known since antiquity. In 300 BC Lu Pu Wei describes crooked legs and hunchbacks and in 180 AD Galen discusses rickets. In 1230 Bartholomeus Anglicus wrote, 'The limbs of the child may easily and soon bend and take diverse shapes. Children's members and limbs are bound with bandages that they be not crooked or evil shaped.' This was the origin of binding infants, creating the symbol of some old Italian children's hospitals. It also stopped children crawling, which was thought to be a throwback to our animal past.

The first scientific descriptions of a vitamin D deficiency, namely rickets, were provided in the seventeenth century by both Dr Daniel Whistler and Professor Francis Glisson. In 1650 Francis Glisson produced a book on rickets, calling it rachitis – from the Greek for spine. Rickets it seems is an old word related to the phrase 'he ricked his ankle', meaning to twist or bend and injure it.

In 1696 a window tax was introduced in Britain when the financially hard-pressed government started taxing properties based on the number of windows. The citizenry responded by bricking up windows and the darker houses are thought to have contributed to an increased incidence of rickets and tuberculosis.

The major breakthrough in understanding rickets was the development in the period 1910–30 of nutrition as an experimental science and the appreciation of vitamins. Vitamin D, however, is a steroid hormone and it is something of a historical accident that it is called Vitamin D. It was in 1919 that Sir

Edward Mellanby, working with dogs that were raised exclusively indoors away from sunlight, devised a diet that allowed him to establish that the bone disease rickets was caused by a deficiency of a trace component in the diet. In 1921 he wrote, 'The action of fats in rickets is due to a vitamin or accessory food factor which they contain, probably identical with the fat-soluble vitamin.' Furthermore, he established that cod liver oil was an excellent antidote. Scientists also observed that when a precursor of vitamin D in the skin was illuminated by sunlight, vitamin D was made. They excised a small portion of skin, irradiated it with ultraviolet light, and then fed it to groups of rachitic rats. The skin that had been irradiated provided absolute protection against rickets, whereas the unirradiated skin provided no protection whatsoever.

Vitamin D is vital to us. It can be absorbed from food by the intestines or it can be produced by the skin when exposed to sunlight. Vitamin D acts as a hormone to regulate the absorption of calcium from the intestine and to regulate the level of calcium and phosphate in the bones.

My grandfather grew up in Birmingham in the English Midlands during the first few decades of the twentieth century, when the city was heavily industrialised. He lived in a terraced house with a back garden that looked down over the railway and canal, and I remember as a young boy many a Sunday afternoon spent there sitting among the long grass of the embankment, in the sunshine, listening to the crickets and the steam trains. He used to tell me that when he was at school, which he left at 14, some of his friends had rickets brought about by bad diet and lack of sunlight. Decades later when I was at school I used to drink a free bottle of milk at morning breaktime, given to all so that we might have calcium to combat rickets. I did not need it but drank it anyway, though I could see some of my friends relied on it more than I. Occasionally there was a child in the school who had appointments at the local hospital to sit in front of a sun-ray lamp. In those days the Sun was not a priority in life among the terraced houses of inner-city Birmingham.

Two years ago, a disturbing report charted the re-emergence of rickets in the Midlands, mainly among those with dark skins, which need longer for UVA rays to reach the vitamin D precursor.

Tanning and skin cancer

However, for some sunlight does not bring depression or deformity but mortal danger. Our skin is our largest and heaviest organ. It is composed of two layers, the epidermis and the dermis. The epidermis is the outermost layer, which principally functions as a protective barrier, while the dermis is the vascularised, innermost layer.

Human skin colour is quite variable around the world, ranging from a very dark brown among some Africans, Australians and Melanesians to a near yellowish pink among some northwest Europeans. The colour is due primarily to the presence of a pigment called melanin, which both light and dark complexioned people have. Two forms are produced – pheomelanin, which is red to yellow in colour, and eumelanin, which is dark brown to black. People with light-complexioned skin mostly produce pheomelanin, while those with dark-coloured skin mostly produce eumelanin. In lighter skin, colour is also affected by red cells in blood flowing close to the skin. To a lesser extent, the colour is affected by the presence of fat under the skin and carotene, a reddish-orange pigment in the skin. Melanin is normally located in the epidermis. It is produced at the base of the epidermis by specialised cells called melanocytes.

Evolution has produced people with darker skin in tropical latitudes, especially in nonforested regions, where the ultraviolet radiation from the Sun is most intense. Melanin acts as a protective shield against ultraviolet radiation, preventing the sunburn damage that could result in DNA changes and, subsequently, melanoma – cancer of the skin.

Tanning is an increase in the number and size of melanin granules due to stimulation by ultraviolet exposure. While skin tanning is often most noticeable on light-complexioned people, even those with very dark brown skin can tan as a result of prolonged exposure to the Sun. However, some northwest Europeans have lost the ability to tan. Their skin burns and peels rather than tans. They are at a distinct disadvantage in tropical and subtropical environments. Not only do they suffer the discomfort of readily burning, but they are at a much higher risk of skin cancer.

It would be harmful if melanin acted as a complete shield. As we have seen, a certain amount of shortwave ultraviolet radiation must penetrate the outer skin layer in order for the body to produce vitamin D. However, too much ultraviolet radiation penetrating the skin may cause the breakdown of folic acid in the body, which can cause anaemia. Pregnant women who are deficient in folic acid are at a higher risk of having babies with neural tube defects. Because folic acid is needed for DNA replication in dividing cells, its absence can have an effect on many body processes, including the production of sperm cells. It may be that the ability to produce melanin was selected for in our early human ancestors because it helped preserve the body's folic acid supply in addition to reducing the chances of developing skin cancer.

People who live in far northern latitudes, where solar radiation is relatively weak most of the year, have an advantage if their skin has little shielding pigmentation. Evolution logically selects for less melanin when ultraviolet radiation is weak. In such an environment, very dark skin is a disadvantage because it can prevent people from producing enough vitamin D, potentially resulting in rickets in children and osteoporosis in adults. Women who had prolonged vitamin D deficiencies as girls have a higher incidence of pelvic deformities that prevent normal delivery of babies.

The Inuit of the American Subarctic are an exception. They have moderately heavy skin pigmentation despite the far northern latitude at which they live. While this is a disadvantage for vitamin D production, they apparently make up for it by eating fish and sea mammal blubber, which are high in vitamin D. In addition, the Inuit have been in the far north for only about 5000 years. This may not have been enough time for significantly lower melanin production to have been selected for by nature.

Such a distribution pattern of human skin colour was predicted by Wilhelm Gloger, a nineteenth-century naturalist. In 1833, he observed that heavily pigmented animals are to be found mostly in hot climates where there is intense sunshine. Conversely, those in cold climates closer to the poles commonly have light pigmentation. But there are exceptions to Gloger's rule in the animal kingdom and they can all be explained by natural selection. In some cases, these are due to the fact that the survival value of having a camouflaged body can be more important than the selective pressures of ultraviolet radiation.

Our distant ancestors may have had light skin and dark hair as they came down from the protection of the trees and made their way out over the savannah and a new way of life. Losing our hair would have made us more vulnerable to UV, so we increased melanin production to compensate. But folic acid in women is destroyed by the longer wavelengths of UV light. So here is the answer: in evolutionary terms we are being pushed two ways at the same time by our adaptation to the rays of the Sun. We need some UV light to make vitamin D, but we don't want so much that all our folic acid is destroyed and we start getting cancers. Our skin colour is an evolutionary balancing act caused by the Sun.

Sunlight can damage the skin, causing premature aging and sometimes cancer. At birth we have a flexible, but tough, protective covering. Our skin is unblemished and wrinkle free, with a smooth texture and a good elastic tone. The surface is covered with fine hair. But with age, all parts of the skin diminish in size and function. It becomes thinner, drier and more fragile. Pigment cells are less active so that the skin tans less easily. Hair becomes finer, thinner and grey. Sunlight injures the skin, particularly the epidermis. The cumulative effects of sun exposure are wrinkling, blotchy pigmentation and roughness.

Sun-damaged skin also becomes less flexible and more easily bruised. Finally, sun damage is the major cause of skin cancer.

There are many genetic disorders that result in the cells of the skin being unable to repair damage caused to them by sunlight. Ultraviolet light from the Sun can penetrate the upper layers of the skin, causing damage to DNA in the nuclei of cells. In everyday life damage to a cell's DNA is common and can be caused by many things, so the body has many cellular repair mechanisms to repair damaged DNA. It is important to do so, as DNA represents a set of instructions for the cell to carry out. Usually those instructions involve the manufacture of a protein important for cell function or for the cell to divide or even die. But if the DNA is damaged, then the instruction can be altered into one causing the cell to grow and multiply without hindrance. In this way a cancer is caused.

There are a range of possible adverse reactions to sunlight. Porphyria is an allergy to sunlight that causes exposed skin to burn, blister and scar. In extreme cases symptoms can include migraines, nausea, vomiting and chronic pain. Some sufferers can even stop breathing. It is not a single disease but a group of at least eight disorders that differ considerably from each other. A common feature in all porphyrias is the accumulation in the body of 'porphyrins' or 'porphyrin precursor'. These are normal chemicals, but they normally do not accumulate in the body. Porphyria is a result of a deficiency of an enzyme used in the production of heme, the red pigment in blood. The missing enzyme inhibits the body's ability to process the light-sensitive porphyrins, which build up in the blood, stool and urine. The symptoms arise mostly from effects on the nervous system or the skin. Those diagnosed with porphyria must protect themselves from the Sun, eat a moderate diet and avoid certain drugs and alcohol, which can exacerbate the condition.

However, the failure to repair UV DNA damage is the most acute and distressing result. There can be a wide range of symptoms: blistering or freckling on minimum sun exposure to the premature aging of skin, lips, eyes, mouth and tongue. There is a significantly increased incidence of cancer in these same areas, with a strong chance of blindness resulting from eye lesions. The DNA damage is cumulative and irreversible and the only possible thing that can be done is to avoid exposure to damaging UV radiation by staying indoors with sunlight blocked out and to use protective clothing, sunscreens and sunglasses. It is sad beyond words never to know the healing rays of the Sun.

Nowadays we work hard to protect ourselves from those rays. Some nutrition experts say that we have gone too far. In addition to protecting against bone diseases such as rickets, sunlight also protects against osteoporosis and vitamin D may lower the risk of type 1 diabetes, multiple sclerosis, rheumatoid arthritis, heart disease and several forms of cancer. But as some dermatologists see it, these benefits are outweighed by the perils of skin cancer.

Where does the balance lie? Sunburn should obviously be avoided, but new research suggests that excessive caution could starve our bodies of essential vitamin D. Likewise, although calcium is available in food, it is difficult for us to get enough this way. For the vast majority of people, the best source is sunlight inducing it to be released in the skin. Nevertheless, there are obvious dangers of excessive Sun exposure, particularly the risk of life-threatening skin melanoma. Deaths in the UK from melanoma rose by 25% in the five years after 1995. About 7000 people a year are diagnosed with melanoma, and for 1500 it is fatal. The UK Government and Cancer Research UK's SunSmart Campaign is determined to get the message across to 'slip, slop and slap' on high-factor sun cream and protective clothing. In the US, 1.3 million new cases of skin cancer were diagnosed in 2003 and in the same year 10,000 people died of the disease.

A recent report, 'Sunlight Robbery', suggests people think that conditions in Britain are similar to those in Australia and we all think we need to cover up, wear hats and sit in the shade, coated in high-factor sunscreen, at the first sight of the Sun, thereby putting ourselves at risk of deficient levels of vitamin D in our bloodstream. One survey suggests that as many as one in four people now do not have the required level of the vitamin in their blood and the cause may be the assumption that any sort of suntan is bad for us. In fact we should respect but not fear the Sun.

There is good evidence that moderate sunlight is rather good for us. In a recent study of four Lancashire towns, Blackpool, which has more hours of sunshine a year than the other three, has 9% fewer deaths from heart disease. There are studies suggesting that moderate exposure to sunlight reduces high blood pressure, because raised blood pressure is linked with lower levels of vitamin D in northern-hemisphere countries. Mounting evidence also suggests that cancers of the breast, prostate, colon and ovaries might be at least partly influenced by vitamin D deficiency. Vitamin D might also promote cancer cell death and prevent the spread of the disease.

Before the industrial revolution, the poor people worked outdoors and got a dark skin from the Sun, while the upper class stayed indoors. So the upper class delighted in showing off a very white skin. But after the Industrial Revolution, the workers stayed indoors and got very light skin, so the upper class decided that tanned skin was more prestigious. It showed that you had the time and the means to spend lots of leisure time outside.

Personally, I prefer pale and interesting.

ROSA VRSINA

SIVE

SOL

EX ADMIRANDO FACVLARVM
& Macularum suarum Phœnomeno VARIVS,

NECNON

Circa centrum suum & axem fixum ab occasu in ortum annua,
circaq.alium axem mobilem ab ortu in occasum conuersione
quasi menstrua, super polos proprios, Libris quatuor
MOBILIS ostensus,

A

CHRISTOPHORO SCHEINER
GERMANO SVEVO, E SOCIETATE IESV.

AD PAVLVM IORDANVM II.
VRSINVM BRACCIANI DVCEM.

BRACCIANI,
Apud Andream Phæum Typographum Ducalem.

Impressio cœpta Anno 1626. finita vero 1630. Id. iunij. Cum licentia superiorum.

The most important book about the Sun for 100 years – the Rosa Ursina
published 1626–30.

CHAPTER TWELVE

Rosa Ursina

Wrapped up warm, we went out in search of sunspots. My then six-year-old son Christopher carried his favourite Christmas present out onto the back lawn and spread its tripod legs on the frozen surface. Together we fixed the homemade cardboard baffles over the front of the telescope and, with Christopher holding a sheet of white card behind the eyepiece, we turned the scope towards the pale wintry sun. When the bright disc, projected onto the card, was brought into focus, there were two dark specks on the Sun. 'Sunspots?' he asked. 'Sunspots,' I replied, 'and each one bigger than the Earth.' 'Wow.'

Most people have heard of sunspots, but few know what they are. Before the invention of the telescope they were rarely seen because only the largest ones can be seen with the unaided eye and then only when the conditions are right, such as thick cloud or mist.

Numerous sightings exist in the historical records. Our friend Anaxagoras saw one in 467 BC and then Theophrastus (374– 287 BC) in the fourth century BC. However, by far the most extensive pre-telescopic records are found in the far east, especially in the official records of the Chinese imperial courts. The oldest records of sunspots are found in the *I Ching*, the Book of Changes, probably the oldest extant Chinese book. It reads 'a *dou* is seen in the Sun' and 'a *mei* is seen in the Sun'. From the context, the words *dou* and *mei* must mean darkening or obscuration.

In the West, the dominating views of Aristotle concerning the perfection of the heavens meant that sunspots were regarded as impossible, so they were ignored or said to be transits of Mercury or Venus across the solar disk. A very large spot seen for no fewer than eight days in 807 was simply interpreted as a passage of Mercury. In fact, as late as 1607 the great astronomer Johannes Kepler, who deduced the laws of planetary motion that bear his name, wanted to observe a predicted transit of Mercury across the Sun's disk, so on the appointed day he projected the Sun's image through a small hole in the roof of his house and did indeed observe a black spot that he took to be Mercury. But had he been able to follow up his observation the next day, he would still have seen the spot. Since Mercury takes only a few hours to cross the Sun's disk during one of its infrequent transits, a fact that Kepler knew well, he would have known that what he observed could not have been Mercury.

One of the most important texts dealing with England in the early Middle Ages is by John of Worcester, author of *Chronicon ex Chronicus*. In 1128 he made a diagram of the Sun containing two large, dark spots. This drawing is wonderful and as far as we know it is the earliest existing drawing of a sunspot. The surrounding text reads:

> On Saturday, 8 December there appeared from the morning right up to the evening two black spheres against the sun. The first was in the upper part and large, the second in the lower and small, and each was directly opposite the other as this diagram shows.

For an observer without a telescope to be able to see not only the spots themselves but their structure implies that these were very large sunspots, rivalling the great group of April 1947. But the story does not end there. According to the latest historical research, shortly after those sunspots were seen the aurora borealis or northern lights were recorded in Korea.

While John of Worcester was noting down his observations, Korean and Chinese astronomers were recording the same. Chinese accounts state that 'there was a black spot within the

Sun' on 22 March 1129, which 'died away' on 14 April. The Chinese accounts for 1127 to 1129 also record recurrent auroral displays: red, white and violet vapours appeared in the southern skies on eight nights over the two-year period.

Large sunspots can be visible to the naked eye under suitable viewing conditions, but it is never a good idea to look at them. I say again that you must never look at the Sun with the unaided eye or with an optical aid under any circumstances. In the past some astronomers thought they could get away with it and damaged their eyesight as a result. We will hear about some of them later.

Telescopes

The next stage in the observation of sunspots came with the invention of the telescope. We cannot be sure who the real inventor was, but we can be certain that it was not the Dutch spectacle maker Hans Lippershey who is often given credit for it. He certainly had a telescope in 1608, but others had been using them before. Lippershey was preceded by someone called Janssen, of whom we know little, who may have been involved in the telescope's development, and there is also an Italian architect named Giambattista della Porta, who died in 1615.

Certainly, knowledge of the magnifying properties of precious stones dates back to antiquity. The Roman writers Pliny and Seneca refer to a lens used by an engraver in Pompeii. There are also curious references to lenses in Japanese folklore. In one story giant ogres with blond and red hair come to rape and plunder Japan with the aid of a tube 'through which you can see a thousand miles'.

In the 1570s Thomas Digges wrote that he 'hath by proportional glasses duely situate in convenient angles, not onely discovered things farre off, read letters, numbered peeces or money and the very coin and superscription thereof.' William Bourne, in a book called *Inventions or Devices* published in 1578, wrote, 'For to see any small thing a great distance from you, it requireith the ayde of two glasses' and 'to see a man four or five miles from you'.

Whatever their origin, in 1608 telescopes suddenly started appearing everywhere in Europe. On 2 October Hans Lippershey applied for a patent on the device. Soon there were two other applications placed before officials in The Hague requesting a patent for the same machine, but by then it was clear that the secret of the telescope was out and the patent requests were turned down. A few months later 'spyglasses' had reached Paris, London, Milan, Venice and Naples.

In the first decade of the seventeenth century, four astronomers more or less simultaneously turned the telescope towards the Sun and saw sunspots. They were Johann Goldsmid (1587–1616, known as Fabricius) in Holland, Thomas Harriot (1560–1621) in England, Galileo Galilei (1564–1642) in Italy and the German Jesuit Christoph Scheiner (1575–1650).

Thomas Harriot

Thomas Harriot lived and worked at Syon House, an imposing mansion situated to the west of London. By 1597 he was engaged in studies of optics and in July 1601 he discovered a law of refraction that all physics students today unjustly know as Snell's law. To Harriot belongs the oldest recorded sunspot observation. On 8 December 1610 he drew the Sun looking through the mists that rise out of the Thames. He wrote in his journal:

Decemb. 8 mane ho. That altitude of the sonne being 7 or 8 degrees. It being a frost & a mist. I saw the sonne in this manner. Instrument 10/1 B. I saw it twise of thrise. once with the right ey & other time with the left. In the space of a minutes time. after the sonne was too cleare.

His entry does not mention the spots explicitly, even though they are clearly indicated on his drawing. After one series of observations about noon in February, 1612, he wrote that 'my sight was after dim for an houre'. He produced nearly 200 drawings of sunspots, made from 1610 to 1612.

You can still see the originals of Harriot's remarkable drawings of the Sun, and also his amazing map of the Moon. Because Harriot was relatively rich, having the patronage of the wealthy, he could afford to use the best inks and the best parchment. So looking at his sunspot drawings today gives one the feeling of looking at a page that was drawn only yesterday. And there are indications of the humour of the man. In asides to his notes it is not uncommon to find comments like 'no observing today, was a-gossiping'. But where are these scientific treasures. On display in the British Library or in the library of the Royal Astronomical Society or the Science Museum in London? No. These priceless scientific documents are held in a box, in a storeroom, in a country house in West Sussex. Harriot was an outstanding scientist who in my view has not been properly appreciated.

Fabricius

Fabricius was also a remarkable astronomer. He first saw a sunspot on 27 February 1611 (9 March on the Gregorian calendar, not yet adopted in East Frisia from where he observed) and shortly thereafter he teamed up with his father to make further observations. His book *De Maculis in Sole Observatis* (On the Spots Observed in the Sun) appeared in the autumn of 1611, but even though it was sold at the Frankfurt book fair it remained unknown to other observers for some time (the fate of many books even today). He correctly interpreted the movement of sunspots from day to day as due to the rotation of the Sun taking them across its face in about two weeks.

The Fabricius son and father team observed sunspots directly through their telescope shortly after sunrise or before sunset, when the Sun is somewhat dimmer than when overhead, and their account confirms the oft-stated caution about observing the Sun this way:

> having adjusted the telescope, we allowed the sun's rays to enter it, at first from the edge only, gradually approaching the centre, until our eyes were accustomed to the force of the rays and we could observe the whole body of the sun. We then saw more distinctly and surely the things I have described [sunspots]. Meanwhile clouds interfered, and also the sun hastening to the meridian destroyed our hopes of longer observations; for indeed it was to be feared that an indiscreet examination of a lower sun would cause great injury to the eyes, for even the weaker rays of the setting or rising sun often inflame the eye with a strange redness, which may last for two days, not without affecting the appearance of objects.

Inevitably they soon started using the technique developed by Johannes Kepler, where an image of the Sun is projected onto a flat surface behind a telescope, the way my son and I did. Fabricius died on 19 March 1616 and his father, who did not pursue the sunspot observations, was killed the following year by an enraged parsoner he had attacked from the pulpit over a stolen goose.

Galileo

It was not long before the professor of mathematics at the University of Padua heard about the telescope. Galileo Galilei said he was 'seized with a desire for the beautiful thing'. At 45 years old Galileo knew an opportunity when he saw one, and he reasoned that the device, two lenses and a tube, could not be all that difficult to construct. If he could present one quickly enough to the Venetian Senate a rich reward or a pension would be his and he certainly needed the money. There was no time to lose. Spyglass salesmen were in Padua and some had plans to go to Venice. Galileo sent word to the Venetian Senate not to accept any rival claims for the instrument and a few weeks later he had made one.

On 25 August 1609 Galileo led a procession of Venetian senators across the Piazza San Marco and up a tower so that they could look across Venice and out to sea, where they saw ships several hours before they could be picked up by the unaided eye. He got his reward, a fivefold increase in salary for life. Giambattista della Porta protested and said he had done such a thing 20 years earlier in a book he wrote about concaves and convexes, but he had failed to see the potential of the telescope. But then, even Galileo failed to realise just how revolutionary it really was. Galileo went a little too far, however, when he sold the councillors of Venice the exclusive right to manufacture the telescope. When they realised this was worthless they asked for their money back.

In October 1609 on a visit to Florence, Galileo showed his telescope to his former pupil Cosimo de' Medici, now the grand duke, and together they looked at the Moon. But the view was poor and Galileo thought that the telescope could be improved. By November it had been and so, with a telescope capable of magnifying up to 20 times tucked under his arm, he walked into his garden in Padua and pointed his so-called perspicillum towards the Moon.

It was one of those rare moments when the Universe changed, when the speculations and prejudices of antiquity fell away. Galileo would call his telescope the 'old discoverer' and it can still be seen at Arcetri near Florence (as indeed can Galileo's preserved finger, which first pointed the way towards the new heavens). He knew that what he was seeing was revolutionary and he rushed into preparing a book. He produced watercolours for it and the *Sidereus Nuncius*, or Starry Messenger, appeared the following March to 'unfold great and wonderful sights ... to the gaze of everyone'.

Galileo's early observations were direct observations of the Sun through the telescope at sunset. He does not describe the procedure, but notes that 'when we look at the brilliant solar disk through the telescope, it appears much brighter than the field which surrounds it'. He describes the sunspots, 'the smallest of them ... when observed through the telescope, can scarcely be perceived and only with fatigue and injury to the eyes'. He mentions viewing at sunset but later he adopted the method of telescopic projection onto a screen so he could observe all day.

Galileo was lucky not to injure his eyes permanently. In the Florence History of Science Museum, his instruments have been measured and tested and the image brightness can be estimated. One 14 power telescope of his would provide an image of the Sun that is 40% as bright as the image on the retina of the unaided eye. A 21 power telescope of his would provide an image 6% of the brightness of that seen by the naked eye. The brightness of the Sun seen through his telescopes would have been further reduced by poor quality, uncoated glass and misalignment. Used at sunset through clouds, it appears that they would not lead to eye injury by focused sunlight. He was fortunate.

Castelli (1578–1643) was a student of Galileo and wrote to him about the projection techniques for observing the Sun, including solar drawings made by projection onto a circle of standard diameter. This is not only safer than viewing with the eye, but it improves the scientific approach to solar observing, by allowing two viewers to see the same image and permitting an accurate tracing of the image to be made. Castelli had begun keeping an accurate record of the movement of sunspots, starting a collaboration with Galileo, and had marked the disk into 15 parts, progressive measurements showing the movement of sunspots. These accurate measurements allowed Galileo to end a controversy by demonstrating that sunspots were at or near the solar surface.

In one of his letters to Christoph Scheiner Galileo writes:

The method of drawing the spots with complete accuracy ... was discovered ... by ... Benedetto Castelli ... Direct the telescope upon the sun ... Having focused and steadied it, expose a flat white sheet of paper about a foot from the concave lens; upon this will fall a circular image of the sun's disk ... The more the paper is moved away from the tube, the larger this image will become, and the better the spots will be depicted ... In order to picture them accurately, I first describe on the paper a circle of the size that best suits me and then, by moving the paper towards or away from the tube I find the exact place where the image of the sun is enlarged to the measure of the circle I have drawn ... if the paper is oblique, the section will be oval and not circular and therefore will not perfectly fit the circumference drawn on the paper.

By tilting the paper the proper position is easily found ... But one must work dextrously, following the movement of the sun and frequently moving the telescope, which must be kept directly on the sun. The correct position may be recognised by looking in the convex lens, where one may see a little luminous circle that is concentric with this lens when the tube is properly pointed toward the sun ... Next one must note that the spots come from the telescope inverted, and reversed from their positions on the sun; that is, from left to right and from top to bottom; for the rays intersect one another inside the tube before coming through the concave lens. But since we draw them on the side of the paper facing the sun, we have the picture opposite to our sight, so that the right-to-left reversal is already effected ... if we merely turn the paper upside down ... we have then only to look through the transparency of the paper against the light, and the spots will be seen precisely as if we were looking directly at the sun.

It is a scene familiar to anyone who regularly observes the Sun through a telescope.

Galileo Galilei played a pivotal role in science and in the story of the Sun. He obtained observational evidence that showed that it was the Sun that was the centre of the solar system and not the Earth, an observation that set him on a collision course with the Catholic Church. The ideas of Copernicus and the observations of Galileo changed our place in the Universe both physically and spiritually and afterwards science and faith went their separate ways.

Christoph Scheiner

While his observations were placing the Sun at its rightful place at the centre of the solar system, Galileo had used his perspicullum to observe sunspots, showing them to a number of people during his triumphant visit to Rome in the spring of 1611. But with all the diversions of fame he did not undertake a regular study of sunspots until April 1612. Scheiner began his study of spots in October 1611 using a telescope with filters of coloured glass. His first tract on the subject, 'Tres Epistolae de Maculis Solaribus Scriptae ad Marcum Welserum' (Three Letters on Solar Spots written to Marc Welser) appeared in January 1612. Welser was a scholar and banker in Augsburg and a patron of local scholars.

Hearing about Galileo's discoveries, Scheiner set out to get a good telescope and became the first astronomer of the telescopic age to establish a solar observatory. Although he was neither the first to observe sunspots nor the first to publish on the subject, his book was the start of a controversy with Galileo over them. Scheiner, a Jesuit mathematician, believed, as Aristotle and the Catholic Church taught, that the Sun was perfect and unblemished and therefore argued that sunspots were satellites of the Sun. As part of a correspondence on the controversy, Welser invited Galileo to comment on Scheiner's writings. Galileo responded with two letters to Welser in which he said that sunspots are on or near the surface of the Sun, that they change shape, that they are often seen to originate on the solar disk and perish there, and that therefore the Sun is not perfect.

Scheiner went on to publish books on atmospheric refraction and the optics of the eye as well as his work on sunspots. When his wealthy patron died on a voyage to Spain in 1624, he went to Rome, where he published his greatest work, the *Rosa Ursina* (1630), the standard work on sunspots for more than a century and so called because Scheiner's patron was the Orsini family, whose crest had a rose and a bear.

In his writings of 1623, Galileo made disparaging remarks about those who had tried to steal his priority in the discovery of celestial phenomena. Although Galileo almost certainly had others in mind, Scheiner interpreted these remarks as being directed against him and therefore devoted the first part of *Rosa Ursina* to an all-out attack on Galileo. It has been said that his enmity towards Galileo was instrumental in starting the Church's attacks against the Florentine in 1633. Scheiner's diatribe against Galileo does, however, not take away from the importance of *Rosa Ursina*, in which Scheiner finally agreed that sunspots are on the Sun's surface or in its atmosphere, that they are often generated and perish there, and that the Sun is therefore not perfect.

Scheiner is generally thought to have been in 1617 the first to construct and use the superior so-called Keplerian telescope, with two convex lenses for an eyepiece, used visually and for projection, although Schyrle de Rheita also claimed to have been first. Scheiner later built a terrestrial telescope with three convex lenses for Duke Maximilian of Tirol and is cited as the first to build an erect image telescope, although here Schyrle de Rheita is a likely predecessor. For his early solar observations, he used blue or green glass filters, observing at sunrise and sunset or through clouds, and quickly scanning the Sun from the edge to the centre.

Scheiner continued sunspot observations for over 15 years, conceiving and building many designs of telescopes for solar projection, which he called the 'heliotropii telioscopici', later contracted to helioscope. He describes them all in *Rosa Ursina*.

His helioscope was the first known equatorially mounted telescope, so that to follow the Sun's path across the sky it had to be moved along only one axis. Several variations of the helioscope are illustrated in *Rosa Ursina*. All of these instruments are built to maintain the drawing easel perpendicular to the telescope for accurate projection. These instruments were used to compile a series of sunspot drawings, combined to create single images showing the path of a sunspot across the surface. These beautiful montages date from 1625–27, allowing the study of sunspot cycles to be traced to that date. About 15 years after the publication of *Rosa Ursina*, the Sun entered the so-called Maunder Minimum, which we shall discuss later. Without spots to observe, the science stalled and Scheiner's work was not superseded for a century.

But the *Rosa Ursina* is much more than a book on the Sun. It includes schematic designs for telescopes and an extraordinary series of illustrations of the correlation between the optics of the eye and the optics of instruments. Scheiner published further books on visual optics and atmospheric optics, none of which have been translated into English.

Johannes Hevelius

It was Johannes Hevelius (1611–87), a wealthy brewer and enthusiastic astronomer, who took the next step in the observation of the Sun. Living in Gdansk (Danzig) in Poland, he used his considerable resources to develop solar telescopes, described in two works, first the mighty *Selenographia* of 1647, which is titled after the Moon but also includes an appendix on sunspots and a chapter on solar observing; and in the *Machina Coelestis*, part one of 1673, Hevelius's elaborate text on his instruments and techniques.

He viewed a solar eclipse in 1630 and another in 1639 and began regularly observing sunspots in 1642. In *Selenographia*, Hevelius presents a value for the average period of solar rotation of 27 days, based on his sunspot observations. They were mostly made with his first helioscope, a modification of Scheiner's instrument. An outer wall was pierced and a socket was mounted to hold a sphere. The ball was fitted with a telescope that could swivel and thereby track an object across the sky. The room was darkened and the telescopic image was projected onto a moveable easel. This surface was held vertically on a bench, the height of which was set to an appropriate level using threaded knobs that engaged threaded vertical supports. The easel would track the Sun by being slowly slid across the horizontal rest by an assistant, and the telescope was linked to the easel so that they moved as a unit. The paper mounted on the easel was not perpendicular to the optical axis of the telescope and the projected image was therefore an oval that changed shape from morning to afternoon. Following the motion of sunspots over time was complicated by the need to correct for the distortion introduced by projection.

In *Machina Coelestis*, Hevelius described the difficulties of drawing a moving object and presented an improved helioscope with an easel that was mounted directly onto an extension of the telescope and thus maintained in a perpendicular orientation to the direction of the Sun. The drawing easel was attached to the framework via a screw-driven rack. Coarse movement of the easel was accomplished by sliding it on a horizontal support and fine tracking was assisted by a pair of screws below the easel.

This elaborate setup was used most notably during the transit of Mercury on 3 May 1661. Published calculations predicted a transit between 1 May and 11 May. Hevelius planned 11 days of continuous observation, if need be. The third day was partly cloudy and at 14:00 the Sun appeared for just a few seconds, with a spot; then at 16:30 the Sun reappeared and the spot had moved. Clouds parted again at 17:00 and 19:30. The image of the Sun on the easel was 80 mm across and Mercury was about 1 mm in diameter. Hevelius used his drawings to estimate the times of first and last contact of the planet as it crossed the face of the Sun and from them deduced the elements of Mercury's orbit. He measured Mercury's apparent diameter at 12 arc seconds – close to the actual value of almost 13 arc seconds.

In Hevelius's *Selenographia* are 26 drawings that accurately show the fainter outer collar of sunspots – the penumbra – and he describes their changes in shape with the rotation of the Sun. He would never have guessed it, but his meticulous work in recording and tracking sunspots was brought to the forefront of solar research some 350 years later when it was realised that the Sun he observed was a rather differently behaved one from the star we see today.

He was elected to the British Royal Society in 1664 and in 1666 was offered the directorship of the newly erected Paris observatory, which he declined and eventually Giovanni Domenico Cassini was appointed. Hevelius's solar observations were published as appendices to his 1647 *Selenographia* and his 1668 *Cometographia*, as well as to his 1679 *Machinae Coelistis*. He used his sunspot observations to determine the solar rotation period to a much better accuracy than his predecessors. He also coined the name *faculae* for the bright regions surrounding sunspots, a name that survives to this day.

Robert Hooke

Robert Hooke by all accounts was not a very nice person, but then neither was his rival Isaac Newton. Hooke had probably more scientific interests than Newton, but he never climbed as far up the peak of scientific achievement and reports say he resented that. He displayed a solar telescope at a meeting of the Royal Society and presented a variety of specialised solar instruments in a remarkable booklet from 1676, 'A Description of Helioscopes', but it was a series of proposals rather than a record of experiments and by the time it was published, Hooke was then off on another tangent. In February of the following year he displayed a box housing a lens of six feet focus and two internal mirrors to contract the optical path into a shortened telescope. Glass mirrors of low reflectivity were used reflecting the Sun's light into the stationary telescope. Evidently Hooke was protective about his shortened telescopes. Four years previously he had engaged Newton in a dispute over which one of them had invented the reflecting telescope. The use of a plane mirror to direct light into a fixed telescope for astronomical purposes seems to have been first proposed by Hooke, at a meeting of the Royal Society. His last recorded scientific investigation, in 1702, was an attempt to measure the solar diameter more accurately.

He proudly wrote of his helioscope:

> which shall so take off the brightness of the Sun, as that the weakest eye may look upon it, at any time, without the least offense. My contrivance is, By often reflecting the Rayes from the surfaces of black Glasses, which are grownd very exactly, flat and very well polished, so to diminish the Radiations; that at length they become as weak and faint as those of the Moon in the twilight, so that one may with ease, and very much pleasure, view, examine and describe the phase of the Sun, and the macula and facula thereof.

John Winthrop

In the eighteenth century progress in our understanding of the Sun was slow because the underlying physics was not understood, but splendid observations continued. In 1739, John Winthrop (1714–79) of Cambridge, MA, became the first astronomer of note in the New World to conduct sunspot observations. His drawings exist as one-page reports in the archives of Harvard University, though they were never published. In 1761, he went on an expedition to St John's, Newfoundland, to observe the transit of Venus across the Sun.

Imaginative, clear thinking and analytical, Winthrop was an unassuming person who it was said could with equal ease discuss philosophical matters or oversee the delivery of milk to the college buttery from his small Cambridge farm. The Reverend Charles Chauncy expressed Bostonians' esteem for Winthrop, remarking that 'none will dispute his being the greatest Mathematician and Philosopher in this Country'. Winthrop's international reputation was established through publications, largely by the Royal Society, on diverse topics such as planetary transits, the nature of comets and earthquakes and the aberration of light.

Unexpectedly noticing a sunspot on the haze-shrouded sun in April 1739, Winthrop wrote in his notebook, 'I plainly saw with my naked eye a very large and remarkable spot.' He studied the phenomenon with Harvard's eight-foot telescope, discovering several additional sunspots, and sketched their appearance. Briefly he added, 'At night a considerable borealis' (northern lights), leaving uncertain whether he surmised the interrelationship of the two phenomena.

The transit of Venus in 1761 gave an unusual opportunity to calculate the solar parallax, essential in determining the distance of the Earth from the Sun. In a ship provided by Governor Francis Bernard and with instruments lent by Harvard, Winthrop sailed to St John's, Newfoundland, to observe the planet's course on 6 June. Bedevilled by swarms of insects and threatening weather, Winthrop and his two student assistants made the only successful observations in North America. Unable to arouse interest in an expedition to Lake Superior to view the 1769 transit and hindered by ill health, Winthrop was limited to calculations from Cambridge.

Alexander Wilson

Alexander Wilson, a professor at Glasgow University, was an accomplished solar observer in the late 1700s. He was responsible for describing what we now call the 'Wilson effect', seen in sunspots as the Sun rotates. Sunspots when viewed near the limb appear as funnel-shaped depressions in the surface. My own undistinguished drawings of sunspots show the effect very clearly. In 1769, Wilson wrote, '... the camera obscura, which both Scheiner and Hevelius often used and which we find greatly extolled by them ... But spots, when seen in this way, have nothing of that distinctness, which is so remarkable and so pleasing, when they are viewed directly through a good telescope armed with an helioscope, or glass properly smoked.'

Wilson broached no criticism of his observations:

> The lameness of the views given in part II may probably proceed ... from our very imperfect knowledge of the vast range of physical causes which obtain in the universe. But ... no doubts ought to arise ... of the spots being themselves what direct observation declares them, namely, excavations in the sun ... as actually demonstrated by competent observations ... it would be a pity not to rescue it from being drawn into the eddy of some treacherous theory, the nature of all which is to sweep into their vortex and finally to precipitate to the bottom every thing which obstructs their impetuous career.

Who did he have in mind, I wonder.

What are sunspots?

But what are sunspots? Their true nature remained a topic of controversy for nearly three centuries. The universally opinionated Galileo proposed, with unusual reservation, that sunspots may perhaps be cloud-like structures in the solar atmosphere. Scheiner believed them to be dense objects embedded in the Sun's luminous atmosphere. In the late eighteenth century William Herschel followed Wilson and suggested that sunspots were opening in the Sun's luminous atmosphere, allowing a view of the underlying, cooler surface of the Sun.

In the notebooks of the English scientist Stephen Gray (1666–1736), on 27 December 1705 there is a reference to a 'flash of lightning' seen near a sunspot. Whatever sunspots were, they held surprises.

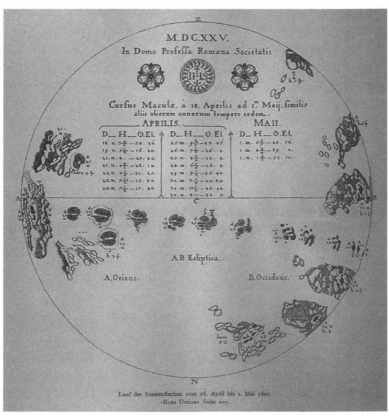

One of the many great sunspot drawings in the Rosa Ursina.

CHAPTER THIRTEEN

The Devil's Jumps

A walk along the three low hills that are called the Devil's Jumps near Churt in the Hampshire–Surrey Hills is pleasant whatever the season. They are so called because a local legend has it that the devil stole a cooking pot nearby and leapt across the countryside in his escape, leaving the three low hills where he jumped across the landscape. Their climb is easy and begins adjacent to a hostelry. The view is magnificent, either in summer over the heathland or across the snow-swept plain northwards to Frensham Little Pond. The summit of the easternmost hill is marked by a curious rock formation composed of fossilised trees lying on their sides that is a source of endless fascination for my youngest daughter, Emily. From it you can look across the other two Devil's Jumps and down to Jumps House, which Tennyson once considered buying and in whose grounds the middle hill stands. It is a place with a solar connection and the scene of a tragedy concerning one Richard Carrington.

The following is an excerpt from 'Description of a Singular Appearance seen in the Sun on September 1, 1859' by Richard C. Carrington, *Monthly Notices of the Royal Astronomical Society*, vol. 20, 13–15, 1860:

> While engaged in the forenoon of Thursday, September 1, in taking my customary observation of the forms and positions of the solar spots, an appearance was witnessed which I believe to be exceedingly rare. The image of the sun's disk was, as usual with me, projected on to a plate of glass coated

with distemper of a pale straw colour and at a distance and under a power which presented a picture of about 11 inches diameter. I had secured diagrams of all the groups and detached spots and was engaged at the time in counting from the chronometer and recording the contacts of the spots with the cross-wires used in the observation, when within the area of the great north group (the size of which had previously excited great remark), two patches of intensely bright and white light broke out ...

My first impression was that by some chance a ray of light had penetrated a hole in the screen attached to the object glass, for the brilliancy was fully equal to that of direct sun-light; but by at once interrupting the current observation and causing the image to move ... I saw I was an unprepared witness of a very different affair. I therefore noted down the time by the chronometer and seeing the outburst to be very rapidly on the increase and being somewhat flurried by the surprise, I hastily ran to call some one to witness the exhibition with me and, on returning within 60 seconds, was mortified to find that it was already much changed and enfeebled. Very shortly afterwards the last trace was gone. In this lapse of 5 minutes, the two patches of light traversed a space of about 35,000 miles.

Carrington had seen an enormous explosion on the solar surface − a rare white light flare. It was not a coincidence that no more than 17 hours later a great disturbance in the Earth's magnetic field took place and aurora were seen as far south as Cuba, as material thrown off the Sun at the time finally reached the Earth.

Richard Carrington belongs in any pantheon of early solar astronomers. In the course of his research, he fixed the 'prime meridian' on the Sun and even today we speak of Carrington longitude and Carrington rotations.

Born in London in 1826, the son of a brewer (another one) and educated at Cambridge, he became an astronomical observer at

Durham University until in 1853 he set up his own observatory at Redhill, Surrey from where he observed the white light flare. As I shall describe in more detail later, in 1843 the German astronomer Heinrich Schwabe had shown that sunspots come and go every 11 years. It was a major discovery and Carrington was appalled that there had been no systematic study of them, so he observed sunspots for seven years, plotting their positions and movements by a method of his own devising. But that was not all he did. His catalogue of 3735 northern-hemisphere stars was so highly regarded that it was printed by the Admiralty at public expense.

Carrington's massive 1863 tome, entitled *Observations of the Spots on the Sun*, ensured that he was recognised worldwide as the British authority on sunspots. He carried out extensive correspondence with sunspot observers across Europe, including Heinrich Schwabe and Rudolf Wolf. Indeed, when Schwabe was awarded the Royal Astronomical Society's Gold Medal in 1857, Carrington personally delivered the medal to the ageing German astronomer and later persuaded him to donate his extensive collection of sunspot drawings to the Society's Archives.

Although Carrington failed to cover a full sunspot cycle, as he originally intended, he reaped a rich harvest from his observations, including the discovery of the Sun's differential rotation (that is, the Sun rotates faster at the equator than the poles), the equator-ward migration of spots during a sunspot cycle (both of these more or less simultaneously and independently discovered by Gustav Spörer in Germany), the determination of the Sun's rotation axis with an accuracy hitherto unprecedented and, of course, observing that white light flare.

Carrington's observations of sunspots showed that the Sun did not rotate in a straightforward manner as if it were a solid body. Its average rotation period is 27 days, but it rotates faster at the equator than at the poles. The rapid development of spectroscopic techniques in the second half of the nineteenth century offered a way to measure the rotation of the Sun by using the Doppler effect to detect the wavelength shift of spectral lines between the approaching and receding solar limb. It was first achieved by Hermann Vogel (1841–1907) in 1871 and a few years after by Charles Young (1834–1908). These results were accurate enough to demonstrate that sunspots rotate at very nearly the same rate as the Sun's visible surface. By the late 1880s Nils Dúner (1839–1914) was carrying out accurate spectroscopic rotational period determinations at latitudes about twice as high as those in which the sunspots usually appeared, demonstrating that the Sun's polar regions rotate about 30% more slowly than its equator.

Interestingly, Christoph Scheiner had already noted in his *Rosa Ursina* that the rotation period inferred from tracking sunspots at different latitudes showed a systematic increase with latitude.

The death of Carrington's father in 1858 forced him to take over the family business, for which he had little interest or flair. In 1865, having fallen into ill health, he sold the family brewery and retired to an isolated spot at Churt, Surrey, where he established a new observatory. An engraving in the *Illustrated London News* of 16 September 1871 shows what a magnificent place it was. On the central Devil's Jump is his observatory beneath which are two tunnels, one horizontal and one vertical, in which to house his clock in a stable environment. But there were not to be many years of productive observations from his new observatory.

He wrote in the *Monthly Notices of the Royal Astronomical Society*:

> Being on a hill I did not want elevation, so I have sunk the observatory below ground, just peeping out over the soil. But I have further sunk a dry well, six feet in diameter, to a depth of forty feet from the centre of the observatory, with a horizontal shaft communicating with the south side of the hill, 166 feet in length, closed with three doorways. This is principally intended for the clock, for I am determined that one clock at least shall be properly mounted, at a position of invariable temperature and in an air-tight case. I propose to reduce the pressure to twenty-seven inches of mercury ... I

hope I shall shortly have the most perfect clock in England, perhaps in the world.

However, it seems that the devil does not take kindly to those who have the audacity to drill holes in his jumps. In August 1869, not long after he moved to Churt, Carrington saw a beautiful woman walking down Regent Street in London and made her acquaintance. She was Rosa Jeffries, some 20 years younger than he was. They married. She could not write and signed the marriage certificate with a cross. However, prior to meeting Carrington she had been living as Rose Rodway, common-law wife of William Rodway, a soldier and by all accounts a nasty sort. He tracked her down to Churt and blackmailed her for money and sex and in 1871, when she resisted, he stabbed her in the arm. She managed to get to a nearby inn and raise the alarm. Rodway was caught and sentenced to 21 years for attempted murder, but died in prison after three years.

Rosa's health suffered following the incident, physically and mentally – it may have been tertiary syphilis, of which there was a lot about. Her wound took a long time to mend. She developed epileptic fits and her mental condition deteriorated badly. Things became very bad and Carrington could not cope, burying himself instead in his work. The morning she was due to be moved to an asylum she was found dead in bed, Carrington had shared the bed but had not noticed! The night before he had given her usual dose of ten grains of chloral hydrate, which she took in 'Hungarian wine'.

An inquest said that she suffocated but did not say how. Dr R Oke Clark, who had been attending Rose, testified that she had been of unsound mind and a recurrence of insanity took place through epileptic fits, of which she had four or five. Rumour spread that Carrington had murdered her. The *Surrey Advertiser* of 20 November 1875 said, 'We learn on the best authority that the conjugal life of the parties was anything but a happy one.' Rosa's death could have been an accident – she was on powerful medication – or it might not have been. Carrington, it was whispered, was too interested in his astronomical studies and she hindered him. Many years later Carrington's own solicitor, J. Alfred Eggar, was in no doubt that the astronomer had murdered his wife, adding that he 'was a man who was wonderfully clever and of exceptional ability, but an infidel'.

After the inquest Carrington vanished for a few days and when he returned to Jumps House the servants had fled. The carriage driver who brought him from Farnham station was the last person to see him alive. A few days later he was found dead, half-dressed, slumped on a mattress in a servant's room with a poultice of tea-leaves tied to his head. The inquest, performed by Dr Clark, said that he died of 'sanguineous apoplexy' or cerebral haemorrhage.

Carrington's will stipulated that he should be buried at a depth of between 10 and 12 feet in his grounds without any service being read over his grave or any memorial erected and that 'after my death neither my chin be shaved nor my shirt be changed'.

Twenty years later, when the region became a haven for the rich and famous who looked for an unspoilt country retreat within a short train ride of London, including people like Alfred Lord Tennyson, George Bernard Shaw, Bertrand Russell, Gerald Manly Hopkins and Sir Arthur Conan Doyle, the novelist Thomas Wright, who had settled in the area, wrote *Ian of the Devil's Jumps*, featuring an astronomer who had an observatory with tunnels. In his works Tennyson wrote, 'What is it all but a trouble of ants in the gleam of a million million of suns?'

Today the observatory is no more, but the horizontal tunnel does remain. Inside it is cool whatever the weather and if you close your eyes and listen carefully, you can almost hear the whirr of the clock drive of Carrington's telescope.

Monitoring sunspots

Sunspots were now being monitored regularly. One avid recorder of their appearance was Warren de la Rue, who was born on the Channel Island of Guernsey on 18 January 1815. Having completed his education in Paris, he entered his father's stationery business. He was one of the first printers to adopt electrotyping and in 1851 he invented the first envelope-making machine and the silver chloride battery. He was attracted to astronomy by the influence of James Nasmyth, an engineer and inventor famous for his development of the steam hammer.

De la Rue constructed in 1850 a 13 inch reflecting telescope, mounted first at Canonbury, later at Cranford, Middlesex, and with its aid executed many drawings of the celestial bodies of singular beauty and fidelity. His chief claim to fame, however, is his pioneering work in astronomical photography. In 1854 he turned his attention to solar physics, taking a photograph of the Sun every day from Kew Gardens Observatory from 1854 to 1858 and then from the Royal Observatory, Greenwich from 1873 to 1882.

The early slow Daguerreotype photographic process was replaced in time by the wet and dry 'collodion' process, in which plates were prepared by an involved method that left them in a layer of collodion, using nitrocellulose impregnated with silver halide salts. With the wet collodion process the plates had to be prepared immediately before exposure and used while still wet. The dry plates were much less sensitive but still suitable for solar photography because of the huge amount of light available. In April 1854 John Herschel wrote to Kew Observatory of the need for daily photographs of the Sun. Kew was at that time a centre for the calibration and testing of instruments to measure the Earth's magnetic field. Kew in turn asked De la Rue, who was active in astrophotography, and in June 1854 he was placed in charge of a major solar photography project. A new specialised instrument for solar photography was its initial goal.

In 1857 De la Rue designed a camera at his private observatory, which he called a photoheliograph. This was in effect a telescopic camera. The first photoheliograph was installed in the King's Observatory at Kew in 1858, but it took two years to solve its many operational problems. It was used until 1874 when it was replaced by a Dallmeyer instrument.

Solar photographs were displayed by De la Rue at a Royal Astronomical Society meeting in 1859, showing faculae and impressive detail. Stereoscopic views were shown, exposed one to three days apart, which were claimed by De la Rue to reveal that the faculae were in the outermost regions of the solar atmosphere.

De la Rue took his photoheliograph to Rivabellosa, Spain for the solar eclipse of 18 July 1860. Since it magnified the image and therefore diminished the Sun's intensity by 64 times, it was not known if the total part of the eclipse would register on film and De la Rue had considerable misgivings about the expedition. At Kew the instrument was on a massive iron mount and a more portable cast iron mounting was made for the expedition. De la Rue devised a prefabricated portable building, half for the telescope and half for the darkroom, with a canvas outer roof that was kept wet to lower the temperature.

At Rivabellosa the sky full of clouds cleared in time, but a reliable pocket chronometer was inexplicably eight minutes fast, so De la Rue's team missed first contact but still managed to take 35 good photographs. At totality, the aperture restriction of the telescope was removed and its full 3.4 inches were used; two photographs taken during totality were later placed in a stereoscope and reportedly showed a spherical Moon in front of the Sun, although the spherical impression would have been illusory. The eclipse was also photographed by Secchi at Desierto de la Palmas, 400 km to the southeast, and when the photographs were compared and found to show the same coronagraphic detail, it was finally demonstrated that prominences – huge streams of gas seen next to the Sun during times of solar eclipses – were really solar and not lunar in origin.

Just before that particular eclipse began, according to De la Rue:

[the] excellent and handy servant Juan ... smoked a piece of glass with a wax lucifer-match ... several pieces for the

by-standers ... the demand soon increased, so much that he was scarcely able to keep pace with it, and at length he became so excited that he threw away the matches in all directions without extinguishing them and some, falling in the standing corn, set it on fire ... a few minutes only could have elapsed before the conflagration would have assumed such dimensions as to be beyond the power of man to control ... the crackling sound and the smell of burning straw drew my attention and ... the fire was got under.

De la Rue is rumoured to have said that in all the scramble for the scientific observations during the eclipse he wished he had not encumbered himself with equipment, and if another chance arose he would 'just look'. I have heard other astronomers say the same.

Using the photohelioscope, Kew director Balfour Stewart photographed the Sun every clear day until 1872, a total of 2778 images over a complete solar cycle. During the last year of operation, on 226 days, 381 solar photographs were exposed, a remarkable record for English skies. In 1873 the instrument was taken to Greenwich, where it began solar photography in April 1874. Although it was soon replaced, it was one of several instruments used in an uninterrupted series of photographs until 1965. It was subsequently donated to the Science Museum of London in 1927, where it is now inventory number 1927–124.

While writing this book I saw the transit of Venus across the face of the Sun, something that occurs so infrequently that no one alive had witnessed it. It was remarkable to see the black dot encroach on the Sun and slowly, over the course of about five hours, move across it. A previous transit of Venus, which occurred in December 1874, motivated the development of a new generation of photohelioscopes, partly to determine the distance to the Sun. On this occasion De la Rue was more relaxed than he had been for the solar eclipse:

No difficulties exist in photographing a transit of Venus ... no strain on the nerves would occur, as in the anxiety

consequent on the desire of rendering available every moment of the short duration of a solar eclipse. All the operations could be conducted with that calm so essential for ... the determination of the Solar Parallax.

Nevertheless, the hoped-for measurement of the Sun's distance using the timing of Venus's transit never worked out. Many of the astronomers who returned from the transit with successful photographs found that they could not be measured with the needed precision, as the exact location of the edge of the limb of the Sun or of Venus was blurred when the photographic silver deposit faded between light and dark. In 1881, an international conference in Paris on planning for the 1882 transit, the second of the pair, recommended against the use of photography and the English and German expeditions did not employ it.

The systematic observations of sunspots in the nineteenth century resulted in two great discoveries about the Sun, both of which have a great bearing on the future of both the Sun and the Earth as we enter an age of global warming. To explain we have to revisit the world of the seventeenth century.

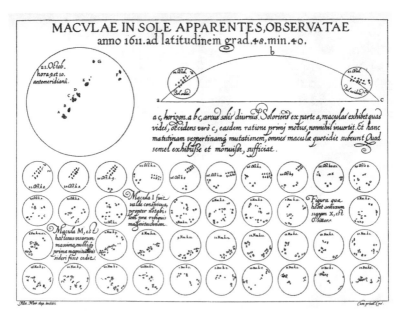

A 17th century solar survey by Jesuit astronomer Christoph Scheiner.

CHAPTER FOURTEEN

The Sun King

When he was fifteen Louis XIV of France danced the part of Apollo at Versailles for the Le Ballet Royal de la Nuit. He had sun emblems on his shoes, his garters, his girdle, tunic and sleeves. From his crown shone the rays of the Sun as he, with appropriate royal dignity, displayed his glitteringly golden costume. Louis XIV went on to the longest reign in European history (1643–1715). During his time he brought absolute monarchy to its height, established a glittering court at Versailles and fought most of the other European countries in four wars. The early part of his reign was dominated by the chief minister Cardinal Mazarin. In the middle period Louis reigned personally and innovatively, but the last years of his personal rule were beset by problems.

He was born at the Royal château in Saint-Germain-en-Laye in 1638. On the death of his father he became king at the age of only five. The regency, confided to his mother Anne of Austria, was marked by a period of rebellion known as the Fronde, led first by the nobility and later by the urban poor. Louis felt both humiliated by arrogant nobles and threatened by the people of Paris and would never forget it. In 1660 he married Maria Theresa, Infanta of Spain. The following year, on the death of his godfather and prime minister, Cardinal Mazarin, the 23-year-old monarch announced that he himself would govern. No one believed him. Yet he insisted on convening a council on a daily basis, from which he excluded grand nobles, surrounding himself instead with ministers who owed him everything.

The first 20 years of the king's personal reign were the most brilliant. With his minister Colbert, he carried out the administrative and financial reorganisation of the kingdom, as well as the development of trade and manufacturing. With the Marquis de Louvois, he reformed the army and achieved many military victories. Finally, he encouraged an extraordinary blossoming of culture: drama (Molière and Racine), music (Lully), architecture, painting, sculpture and all the sciences (including the founding of the royal academies). These accomplishments would be depicted on the ceiling of his grand palace at Versailles.

Louis chose the Sun as his emblem. It was associated with Apollo, god of peace and arts, and was also seen as the heavenly body that gave life to all things, regulating everything as it rose and set. Like Apollo, Louis XIV brought peace, was a patron of the arts and dispensed his bounty. Throughout Versailles the decoration combines images and attributes of Apollo (laurel, lyre, tripod) with the king's portraits and emblems (the double LL, the royal crown, the sceptre and the hand of justice). The path of the Sun is also traced in the layout of the gardens.

The story of Apollo governs the statues and fountains along the main axis of the gardens at Versailles. The lake at the western end is adorned with Apollo rising with his four-horsed chariot from the waves at dawn, beginning his daily path. Nearer to the château is Apollo's mother, Latona, at the end of the central fountain. The cycle terminated in the Grotto of Thetis, where the north wing now stands; it showed the god being tended by nymphs in the evening, as depicted in François Girardon's masterpiece.

Missing sunspots

Something peculiar happened to the Sun during the reign of the 'Sun King'. As we have intimated, but now must cover in more detail, in 1826 the German amateur astronomer and pharmacist Samuel Heinrich Schwabe (1789–1875), set himself the task of discovering intra-mercurial planets, whose existence had been conjectured for centuries. His first telescope had been won in a lottery, but he later bought a more powerful one from Joseph von Fraunhofer. Like many before him he realised that his best chances of detecting such planets lay when they transited the Sun, but the main difficulty was the problem of confusing such planets with small sunspots.

So Schwabe began meticulously recording the position of any sunspot visible on the solar disk on any day the weather would allow. In 1843, after 17 years of observation and having sold the family business, he had not found a single intra-mercurial planet, but had discovered something else. He found a cyclic increase and decrease over time of the number of sunspots visible on the Sun, a solar cycle with a period that he originally estimated to be 10 years.

At first many did not believe him but in 1852, within a year of the publication of Schwabe's results in a book called *Kosmos*, the Irish astronomer and explorer Edward Sabine (1788–1883) announced that the sunspot cycle period was the same as that of the fluctuations in the Earth's magnetic field, for which reliable data had been accumulated since the mid-1830s. But how could such a cycle have gone unnoticed? Slowly Schwabe's discovery gained recognition and others wondered if the cycle could be traced farther back in the past using historical observations. Rudolf Wolf (1816–93), undertook the daunting task of comparing sunspot observations carried out by many different astronomers using various instruments and observing techniques.

Wolf's interest in sunspots had started when he saw a particularly large group in December 1847. Thereafter he was to observe the sun almost daily for 46 years. To help him he defined the relative sunspot number, which is still in use by many astronomers today. By 1868 he had a more or less reliable sunspot number back to 1745, and pushed his reconstruction all the way back to 1610, although the paucity of data effectively rendered these older determinations far less reliable. Wolf was the first to note the possible existence in the sunspot record of a longer modulation period of about 55 years.

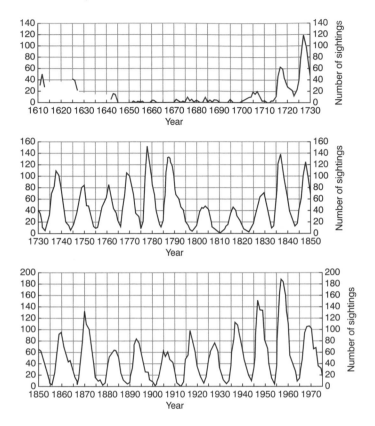

Figure of historical sunspot observations.

There was something odd occurring in the accumulating data of historical sunspot observations. Johannes Hevelius noted the presence of numerous spots in 1644 and the French Jesuit Jean Picard (1620–82) and Jean Dominique Cassini (1625–1712) at the Observatoire de Paris (newly founded by Louis XIV) saw one in 1671. But Wolf noted that very few sunspots were observed from about 1645 to 1715, and when they were seen their presence was considered a noteworthy event by active astronomers. The dearth of sunspots seemed to be a real effect rather than the consequence of a lack of diligent observers. A simultaneous decrease in auroral counts further suggested that solar activity was greatly reduced during this time period.

Edward Walter Maunder

It was the English astronomer at the Royal Observatory, Greenwich, Edward Walter Maunder, who first drew the threads of the mystery of the missing sunspots together. Born in St Pancras, London, he grew up amidst the debates about Darwinism, during a time of automatic patriotism and poverty. It seems that he had an interest in astronomy from a very early age. He later wrote about some sunspots he had seen when he was 14 years old:

> In February 1866, as I was returning home from school one evening, I saw the sun, low down in the west, shining red through the mist. The sun was dim and red enough for me to look at him without blinking and I saw plainly on him a round black spot ... It was the first time I had ever seen anything on the surface of the sun ... the next time that I saw the sun lying low and red in the west, I saw the black spot ... again, but it had changed its place, and was now much further from the centre of the sun. Two or three days later it was gone.

In 1872 Maunder entered King's College, London as an 'occasional student' taking chemistry, mathematics and natural philosophy – what we call physics today. He also found work in a London bank. But later that year he took an entrance exam to join the Royal Observatory at Greenwich. The Astronomer Royal, Sir George Biddell Airy, needed some more staff. Not that he was very kind to some of those who worked for him. In those days the observatory hired children who were good at maths as 'human computers'. But after a few years, presumably when they were regarded to be past their best, they were discharged. Maunder went to work for Airy but did not like him: 'despotic in the extreme' he wrote.

In those days the Greenwich Observatory was a poor performer and many people preferred to work at the Kew Observatory on the other side of London, which was better equipped. The Greenwich Observatory was dominated by the needs of the Admiralty, but despite this in 1872 Airy requested to carry out sunspot and solar spectroscopic work. Maunder was appointed photographic and spectroscopic assistant at £200–300 a year.

From about 1877 he started measuring sunspot areas and the size of faculae, collecting data for his 'butterfly diagram', which showed the equator-ward motion of sunspots during the solar cycle that had been seen by Carrington. It is still a magnificent diagram, based on 9000 photographs of the sun and 5000 separate groups of sunspots over 30 years. It encapsulates so much that is known, and unknown, about the Sun. But Maunder's magnificent chart no longer resides at Greenwich, having left a long time ago.

In 1943, while London was being bombed nightly by the Germans, Maunder's second wife Annie, who outlived him, got a letter from a friend named Stephen Ionides. Born in London, Ionides had led an adventurous life as an engineer and miner in England, Australia, Mexico and the American West before settling in Denver in the United States. Ionides's hobby was science history and he asked Annie for one of her and her husband's drawings. To save the butterfly chart from destruction in the Blitz, she gave it to Ionides. Later he gave it to the Harvard College Observatory. The framed diagram remains on display today at the scion of that organisation, the High Altitude Observatory of the National Center for Atmospheric Research, in Boulder, Colorado.

Maunder also had much to say about the vexed question of the connection between solar activity and fluctuations in the Earth's magnetic field. Despite Sabine having shown that the 11-year sunspot cycle signal shows up in the fluctuations of the Earth's magnetic field, there were many scientists who did not believe such a magnetic connection was possible, among them some very big guns indeed. The fluctuation in the compass needle seen by Carrington the day after his white light flare did not seem to impress them.

Maunder was convinced that there was a connection between the Sun's surface activity and fluctuations observed in the Earth's magnetic field. With his wife, he later wrote:

> In November of 1882, a monster sunspot easily visible to the naked eye, crossed the sun and when it was about halfway across, on November 17th, a very violent magnetic storm, as these agitations of the magnetic needle are called, occurred ... ten years later, in February 1892, a still greater spot ... appeared upon the sun, and when it had passed a little to the west of the sun's centre ... a still more violent magnetic storm occurred than in 1882. This great spot passed off the sun, and returning to the eastern edge, again crossed the sun's disc. When it arrived at the same distance from the centre of the sun, there suddenly broke out again upon the earth a great magnetic storm. Eleven years later, in October 1903, yet another giant sunspot appeared ... and there was a magnetic storm but not a violent one ... but a fortnight later when an important, but smaller, spot had got into the central position of the sun's disc, a magnetic storm burst suddenly ... the most violent that has been experienced in the memory of man.

The mighty physicist William Thompson – Lord Kelvin, who had achieved so much – did not believe that there was a connection, just as he did not believe in Darwinism. He had done calculations that proved that the Sun could not possibly be responsible. Nevertheless, Maunder could see a flaw in Lord Kelvin's calculation. He had assumed that magnetic waves came out of the Sun in all directions – but what if, Maunder suggested, they came out in certain preferred directions? That way they would not be diluted with distance so much. While Maunder was right, Lord Kelvin was influential and his views held sway, setting back the study of solar-terrestrial relations by decades. The question of the missing sunspots fascinated Maunder. Years before, William Herschel, discoverer of Uranus in 1781, had noticed something peculiar, pointing out the lack of sunspots in five irregular periods, 1650–70, 1676–84, 1686–88,1695–1700 and 1710–13. In 1801 he introduced the idea of more transient climate connections when he pointed to periods in the seventeenth century, ranging from two decades to a few years, when hardly any sunspots had been observed. He pointed out that during those periods the price of wheat had been high, presumably reflecting spells of drought.

For Maunder the threads came together with an American scientist's work that linked the missing sunspots with an effect on Earth. Andrew Ellicott Douglass was a remarkable scientist who fell out with his boss Percival Lowell, the great advocate of artificial canals on Mars, because he rightly suspected that these were not real. After he had been treated disgracefully by Lowell, he eventually made a new career almost founding the new science of dendrochronology – the study of tree rings, which were thinner in dry years. Douglass announced some remarkable coincidences between the sunspot cycle and rings in trees after studying beams from old buildings as well as Sequoias and other long-lived trees. He could see evidence for a prolonged dry and cold spell in the seventeenth century. He was ahead of his time: the value of tree rings for climate study was not firmly established until the 1960s.

In 1922 Maunder read a letter from Douglass to a meeting of the British Astronomical Association:

Sequoias show strongly the flattening of the curve from 1670 or 1680 to 1727. Again, taking the evidence as a whole, it seems likely that the sunspot cycle has been operating since 1400 AD with some possible interference for a considerable interval before the end of the 17th century.

Maunder died in 1928 and his work did not receive the attention it deserved, lying half-hidden – as did the lack of sunspots in the seventeenth century – for almost 50 years.

Jack Eddy

It was the American solar physicist Jack Eddy who rediscovered Maunder's work, but he may not have done so had he not lost his job. Born in a small town in southeastern Nebraska, he went to the Naval Academy in Annapolis in 1949 where there was a lot of engineering, and particularly marine engineering. He served four years as a naval officer in the Korean War and then left the navy and was accepted in the graduate school of the University of Colorado to study mathematics.

At that time, in the late 1950s, astronomers were trying to find new ways to observe the Sun's outer atmosphere – the corona – with the help of the coronagraph, which helped it be observable without having to wait for a solar eclipse. High up in the mountains at Climax, Colorado, at an elevation of about 11,000 feet, was an observatory whose principal instrument was a large coronagraph. Eddy was trying to observe the corona better by going to even higher altitudes to see if it were possible to photograph the corona outside of an eclipse with balloons and aircraft, work that eventually led to satellite astronomy and orbital coronagraphs.

A colleague told him of the work of Walter Maunder 100 years before, and his thinking that there was a prolonged period of time in the 1600s when the Sun wasn't so active, 'That really piqued my curiosity,' Eddy said later. 'Something had jolted my professional life, which was probably the worst and maybe also the best thing that had ever happened to me.' In 1973 he had been at the High Altitude Observatory for almost 10 years when there was a major cutback in funding and Eddy was selected to be let go.

He faced an uphill battle convincing his colleagues about the reality of what was called the Maunder Minimum, as it leaned entirely on accounts from so long ago. But he looked for other evidence and found it, as we shall explore later. Because Eddy had been trained in astrogeophysics and knew something of the other ways in which the Sun affects the Earth, he looked hard at historical records of aurorae. He also got acquainted with the Laboratory of Tree Ring Research in Tucson. Carbon-14 data from trees clearly showed a pattern of slow growth at the same period, indicating that at the same time as the sunspots vanished the Earth caught a little cold.

The lack of sunspots during the time of the Sun King was real and there was a connection between the Sun's lack of spots and the Earth's weather. Understanding this connection is crucial for man to flourish on Earth, as we shall later see.

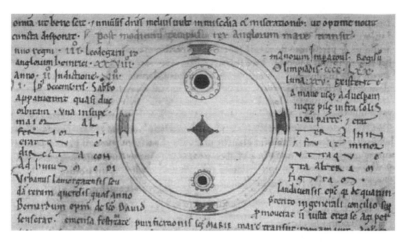

The first known drawing of a sunspot – 8 December 1128 – by John of Worcester.

CHAPTER FIFTEEN

A single ray of light

*I*n his *Treatise on Painting* written in the 1490s Leonardo da Vinci wrote,

> If you mean that the proximity of one colour should give beauty to another that terminates near it, observe the rays of the sun in the composition of the rainbow, the colours of which are generated by the falling rain, when each drop in its descent takes every colour of the bow.

In 1637 philosopher René Descartes wrote,

> A single ray of light has a pathetic repertoire, limited to bending and bouncing (into water, glass or air, and from mirrors). But when rays are put together into a family – sunlight, for example – the possibilities get dramatically richer. This is because a family of rays has the holistic property, not inherent in any individual ray.

The rainbow has always captivated us and it has enchanted poets and scientists alike. 'We have not the reverent feeling for the rainbow that a savage has, because we know how it is made. We have lost as much as we gained by prying into that matter.' So thought Mark Twain.

One man who became important in the study of the Sun was also captivated by the spectrum and built on the work on rainbows done by Newton. He was George Ellery Hale. He revolutionised the study of sunspots and, indeed, the whole of astronomy as well. He showed that sunspots were magnetic phenomena and thereby provided a crucial insight into the energies that influence phenomena on the Sun's surface. He also moved astronomy out of the rich, enthusiastic amateur era into one of industry and professional research.

George Ellery Hale was born in Chicago on 29 June 1868, an only child and heir to his family's considerable fortune. When he was 13 years old he received *Cassell's Book of Sports and Pastimes* and 'found a description of the spectrum, that beautiful rainbow of light first seen through a prism ... which provides a key to the understanding of the physical universe.' He wrote in his notes:

I cannot fix the date of my first interest in astronomy, but it must have been when I was 13 or 14 years old. I built a telescope, but as I used a large single lens, the images were not good ... I learned of a second-hand Clark refractor, of 4-inches aperture. This was purchased by my father and I mounted it on the roof of the house.

I attached a plateholder to the telescope and photo-graphed a partial eclipse of the Sun. I also began to observe Sunspots and made drawings of them.

My father, always ready to encourage serious efforts, enabled me to buy a small spectrometer. This had a single prism and I lost no time in fitting it with a small plane grating ... Nothing could exceed my enthusiasm in observing the solar spectrum and in measuring the principal lines. I bought Lockyer's Studies in Spectrum Analysis and began the observation of flame and spark spectra and their comparison with the spectrum of the Sun. At last I had found my true course and I have held to it ever since.

It was Hale's 1889 invention of the spectroheliograph that would bring him worldwide fame and revolutionise the study of the surface of the Sun. He got the idea while riding a southbound cable car along Cottage Grove Avenue in Chicago. He was thinking about a recent visit he had made to the Halsted Observatory at Princeton. Charles A. Young, its director, had been trying to photograph a prominence in full sunlight with a small spectrograph, but too wide a range of wavelengths were being recorded to get a picture of real value. All Young got was an outline, with no details. Staring at a white picket fence while the cable car moved along, Hale got an idea that the photographic plate should move along with the Sun while the telescope remained stationary. Refining the concept to use two spectrograph slits, Hale saw that a particular spectral line could be isolated and photographed.

In an article published in the *Beacon*, a small Chicago journal, in July 1889, Hale described his work alongside the experiments of Kirchhoff and Bunsen with prisms, which revealed much about the solar spectrum. He recognised that a new astronomy was being born.

He was hired by the University of Chicago in 1892 on condition that the university build a large observatory for him. He had heard of a 40-inch lens that was available and he persuaded Charles Yerkes, a wealthy businessman in Chicago, to pay for an observatory and telescope. Yerkes was a scoundrel who had made millions in Chicago financing the elevated tracks, its overground cable system, and the Peoples Gas Light and Coke Company. Yerkes agreed to pay for the 40-inch lens, telescope and observatory. Hale was to be the Director of Yerkes Observatory, soon the largest astrophysical laboratory in the world, and he was but 24 years old. It would never happen today, unfortunately!

Hale used his spectroheliograph to observe two large flares similar to the one seen by Carrington, which were followed 19.5 and 30 hours later, respectively, by magnetic storms. Later he tuned his instrument to spectral lines known to change when produced in strong magnetic fields, which led in 1908 to the first unambiguous demonstration that sunspots are the seats of strong magnetic fields. It was a key observation, standing alongside

Newton's prism, Herschel's spectrum and Schwabe's sunspot cycle in opening a new door on solar understanding. Not only was this the first detection of a magnetic field outside the Earth, but the inferred magnetic field strength, 3000 Gauss, was over 1000 times greater than the Earth's magnetic field.

In the following decade Hale and his colleagues went on to show that large sunspot pairs almost always show the same magnetic polarity pattern in each solar hemisphere, that they had opposite polarity patterns between the North and South solar hemispheres, and that these polarity patterns are reversed from one sunspot cycle to the next, indicating that the sun's magnetic cycle has a period of twice the 11-year sunspot cycle period. But such insights came at a price.

Two days after graduating in June 1890, Hale married Evelina Conklin at her home in Brooklyn. A honeymoon trip out West included a visit to the Lick Observatory on Mount Hamilton, near San Jose, California. They returned to Chicago where they lived with Hale's parents. Occasionally Evelina would sit with Hale in the observatory to break the monotony of her day, yearning for more gaiety in her life, since Hale's mother suffered from migraines and kept the house dark and quiet. The sombreness vexed Evelina, but she tried to fit in. Although concerned about his wife's feelings, Hale spent every available moment in his laboratory. When Evelina hinted that she wanted to live alone with Hale, his mother told her that this would cause embarrassment to the family. Evelina let the idea drop and continued 'to endure existence in that forlorn house'.

Although he never completed a doctorate, in 1892 Hale was appointed professor of astronomy at the University of Chicago (without salary for the first three years) and launched in a lifelong campaign to fund and build ever better astronomical observatories. After the Yerkes Observatory Hale founded the Mount Wilson Observatory and persuaded the Carnegie Foundation to fund its 60-inch reflector. He convinced John D. Hooker and the Carnegie Foundation to build the even larger 100-inch telescope there.

Overworked and suffering from recurrent episodes of depression, he resigned as director of Mount Wilson Observatory in 1923 and retired from the active scientific research scene in the following years, leaving when he was arguably at the height of his career. His crowning achievement was convincing the Rockefeller Foundation to fund the building of the Palomar Observatory and its 200-inch telescope. Hale did not live to see the telescope's completion, but the instrument was named after him to honour his extraordinary legacy to astronomy. Before he died he wrote,

Starlight is falling on every square mile of the earth's surface and the best we can do at present is to gather up and concentrate the rays that strike an area 100 inches in diameter.

Fascinating insights they were.

But how does the Sun work? Where does it get its power?

Galileo's rival Christoph Scheiner – his Rosa Ursina was a landmark solar treatise.

Go find a hotter place

While delivering a lecture on astronomy in the 1920s, the English astronomer Sir Arthur Eddington gave a brief overview of the early theories of the Universe. Among others, he mentioned the Indian belief that the world rested on the back of a giant turtle, adding that it was not a particularly useful model as it failed to explain what the turtle itself was resting on. Following the lecture he was approached by an elderly lady. 'You are very clever, young man, very clever,' she said, 'but there is something you do not understand about Indian cosmology: it's turtles all the way down!' It's an amusing story that found its way onto the first page of Stephen Hawking's *A Brief History of Time*, though Hawking incorrectly says the scientist might have been Bertrand Russell.

Sir Arthur Stanley Eddington was a slender, nervous-looking man, of whom it was said that he really looked like his photographic studio portrait with all its formality and stiffness. He wrote with great clarity: even today his works, though dated, are well worth reading and he has a turn of phrase that can astonish. However, it seems that he did not lecture with the talent that he used to put pen to paper. He 'began in mid-sentence and stopped at the end of the hour without any full stops in between', as his lecturing style was described. He once wrote that he hoped one day to understand 'so simple a thing as a star'.

Eddington was one of the key figures in our understanding of the Sun and he had a profound faith in science. He once said that the existence of stars could be predicted. Imagine living on a completely cloudy planet with no view of the stars. From experiments carried out in the laboratory on masses of atoms, competent physicists could predict that a mass of gas of a certain size would be hot enough to radiate light. Imagine, he said, armed with this prediction, if the scientists in question were able to take off in a rocket and travel above the atmosphere and for the first time see the Universe above, and all the radiant stars, just as predicted.

The young Stanley, as his family called him, showed early mathematical promise. Before he could read he had learnt the 24 times 24 multiplication table and he was fascinated by the stars. At age ten he was loaned a three-inch telescope and he would give lectures on astronomy to the devoted household servant. His schooling at the Brynmelyn School in Weston-super-Mare fostered his academic ambitions and a scholarship in 1898 allowed him to enter Owen's College, a nonconformist foundation that by that time had become Victoria University and in 1902 became Manchester University. Eddington was just 16 years old and his genius was by no means confined to science. In his second year he topped the class in Latin and English History as well as Mathematics and Mechanics. He completed his degree in physics in 1902 and left for Trinity College, Cambridge, at the age of 19.

He began a research project in the Cavendish Laboratory but he soon gave up the project. His next research in mathematics was no more successful than his work in physics. Had he been starting off in science today he would have never got very far with such an early track record. With something of a question mark over his scientific potential, he made the move to astronomy with his appointment to a post at the Royal Observatory at Greenwich. He became involved in a project that had been underway since 1900, when photographic plates of the asteroid Eros had been taken over the period of a year. Eddington's first task was to complete the analysis of the photographs to determine an accurate value for the solar parallax, which was the key to understanding the distance to our Sun and to other stars. Things started to go better for him. He was a Smith's prize winner for an essay on the motions of stars in 1907 and he was awarded a Trinity College Fellowship.

George Darwin, a son of Charles Darwin and Plumian professor of astronomy at Cambridge, died in December 1912 and the following year Eddington was appointed to fill his post. There were two chairs of astronomy at Cambridge, the other being the Lowndean chair. Originally the Plumian chair covered the experimental side of the subject, while the Lowndean chair covered the theoretical side. Although this distinction had become somewhat blurred over the years, the appointment of Eddington was certainly seen as an appointment in observational astronomy. However, the holder of the Lowndean chair died towards the end of 1913 and Eddington became director of the Cambridge Observatory. In doing so he effectively took over responsibility for both theoretical and experimental astronomy at Cambridge.

Shortly after taking up his role of leading astronomy research at Cambridge, World War I broke out. Eddington's parents had been quakers and, as a conscientious objector, he avoided active war service and was able to continue his research at Cambridge during the war years of 1914–18.

At the same time, Albert Einstein was saying some remarkable things about gravity, claiming that it was the result of curved space. He predicted that a beam of light passing close to the Sun would be deflected slightly due to the curved space around the Sun. Seen from the Earth, a star would appear to be in a slightly different position if the Sun were close by along the same line of sight. The problem was that it was not possible to see stars close to the Sun, except during an eclipse.

To see the effect, Eddington led an eclipse expedition to Principe Island in West Africa. He sailed from England in March 1919 and by mid-May had his instruments set up. The eclipse was due to occur at two o'clock in the afternoon of 29 May, but that morning there was a storm with heavy rain. Eddington wrote:

> The rain stopped about noon and about 1.30 ... we began
> to get a glimpse of the sun. We had to carry out our

photographs in faith. I did not see the eclipse, being too busy changing plates, except for one glance to make sure that it had begun and another half-way through to see how much cloud there was. We took sixteen photographs. They are all good of the sun, showing a very remarkable prominence; but the cloud has interfered with the star images. The last few photographs show a few images which I hope will give us what we need.

He remained on Principe Island to develop the photographs and to measure the deviation in stellar positions. The cloud made the plates of poor quality and hard to measure, but on 3 June Eddington recorded in his notebook: 'one plate I measured gave a result agreeing with Einstein.' Indeed, the results from the Africa expedition provided the first confirmation of Einstein's theory that gravity will bend the path of light when it passes near a massive star. Afterwards Eddington wrote, in a parody of the *Rubaiyat of Omar Khayyam*:

Oh leave the Wise our measures to collate.
One thing at least is certain, light has weight.
One thing is certain and the rest debate.
Light rays, when near the Sun, do not go straight!

In November 1919, shortly after Einstein's theory of relativity was confirmed by observations from Principe, Eddington was approached by Ludwig Silberstein at a joint meeting of the Royal Society and the Royal Astronomical Society.

'Professor Eddington,' Silberstein declared, 'you must be one of three persons in the world who understands general relativity.' When Eddington demurred, Silberstein continued, 'Don't be modest, Eddington.'

'On the contrary,' Eddington replied. 'I am trying to think who the third person is!' Indeed, Einstein himself said that this work was the finest presentation of the subject in any language.

One day, one of Albert Einstein's assistants expressed his joy that Eddington's results had confirmed the general theory of relativity. 'But I knew that the theory was correct,' Einstein calmly remarked. The assistant then asked what he would have done had his predictions not been confirmed. 'Then,' Einstein replied, 'I would have felt sorry for our dear Lord – the theory is correct.'

How old is the Sun?

Eddington was also interested in what happened inside a star and where it got its energy from. However, before we return to him we have to consider how old the Sun is.

One of the parameters that scientists wanted to measure about the Sun is how much heat it deposits on the Earth. This is called the solar constant, and if you measure it you can work out the Sun's total energy output. The solar constant is a measure of the Sun's luminosity and is defined by convention as the amount of energy incident per second on one square metre of the outer terrestrial atmosphere, when the Earth is at a distance of one astronomical unit (149,598,500 km) from the Sun.

The first attempts at a direct measurement were carried out independently and more or less simultaneously by the French physicist Claude Pouillet (1790–1868) and John Herschel. Although they used different methods their underlying principles were the same: a known mass of water is exposed to sunlight for a fixed period of time, and the accompanying rise in temperature recorded. The energy input rate from sunlight is then readily calculated, knowing the heat capacity of water. However, their value for the solar constant was about half the accepted modern value of 1367 ± 4 Watts per square metre, because they failed to account for absorption of the Sun's light by the Earth's atmosphere.

The American scientist Samuel Langley (1834–1906) carried out the most elaborate attempt to determine the solar constant at the time, during an expedition to Mt Whitney, California, in July 1881. Using his recently invented bolometer (an instrument based on the varying electrical resistivity of metals with temperature), as well as other instruments, Langley conducted measurements at different wavelengths and at different altitudes, demonstrating the strong variation with wavelength of the absorption by the Earth's atmosphere. However, the solar constant value he calculated at the time, 2903 Watts per square metre, is nearly a factor of two larger than the modern value.

Joseph Stefan (1835–93), a professor of mathematics at the University of Vienna and the first person to obtain a sensible value for the temperature of the Sun, believed that there was something systematic about the ability of hot objects to give off energy. His pupil Ludwig Boltzmann (1844–1906) also developed the theory. It became the Stefan–Boltzman law and it relates the energy given off by a body to its temperature.

Stefan's work came to the attention of French physicist Jules L. G. Violle (1841–1923), who had a longstanding interest in wanting to know the Sun's temperature. Violle had to determine how much of the Sun's heat was lost as it passed through the Earth's atmosphere. To do this he climbed Mont Blanc and made measurements, while his assistant made similar ones at the same time from below. When Stefan saw Violle's figures he was able to arrive at the correct figure of 6000 degrees Celsius.

So the Sun is a ball of gas with a surface temperature of 6000 degrees. But how did it get so hot, and how long will it last?

The energy source of the Sun was believed by many nineteenth-century physicists to be gravity. In a lecture in 1854 Hermann von Helmholtz, a German professor, suggested that the origin of the Sun's enormous radiated energy is the gravitational contraction of a large mass; a proposal made somewhat earlier, in the 1840s, by J. R. Mayer and J. J. Waterson. It was simple: compress something and it gets hot.

In 1859 Charles Darwin, in the first edition of *On the Origin of the Species by Natural Selection*, made a crude estimate of the age of the Earth and therefore a lower estimate for the age of the Sun, by considering how long it would take erosion occurring at the current observed rate to wash away the Weald, a great valley that stretches between the North and South Downs across the south of England. His estimate for the 'denudation of the Weald' was in the range of 300 million years. This was pleasing to Darwin because it was long enough for natural selection to have produced the wide range of species that exist on Earth.

Firmly opposed to Darwinian natural selection was the familiar Lord Kelvin, a professor at the University of Glasgow and one of the great physicists of the nineteenth century despite his objections to evolution. He formulated the formidable second law of thermodynamics and set up the absolute temperature scale, which was subsequently named the Kelvin scale. The second law of thermodynamics states that heat naturally flows from a hotter to a colder body, not the opposite. Because of this Kelvin knew that the Sun and the Earth must get colder unless there is another energy source and that eventually the Earth will become too cold to support life. He believed that the Sun's luminosity was produced by the conversion of gravitational energy into heat and in a lecture in 1854 he suggested that the Sun's heat might be produced by the impact of meteors falling onto its surface. A few years later he wrote:

That some form of the meteoric theory is certainly the true and complete explanation of solar heat can scarcely be doubted when the following reasons are considered: (1) No other natural explanation, except by chemical action, can be conceived. (2) The chemical theory is quite insufficient, because the most energetic chemical action we know, taking place between substances amounting to the whole sun's mass, would only generate about 3,000 years' heat. (3) There is no difficulty in accounting for 20,000,000 years' heat by the meteoric theory.

So there was a situation whereby the physicists believed that the Sun was at most a few tens of million years old, but many geologists and biologists considered that it must have been shining for at least several hundreds of millions of years in order to account for geological changes and the evolution of living things, both of which depend critically on energy from the Sun. However, Charles Darwin was so shaken by the power of Kelvin's analysis and the authority of his theoretical expertise that in the last editions of *On the Origin of the Species* he eliminated all mention of specific time scales.

Nevertheless, Lord Kelvin was wrong and the geologists and evolutionary biologists were right. We now know from the radioactive dating of meteorites that our Sun is at least 4.6 billion years old. Kelvin and his colleagues calculated a maximum age for the Sun that was far too short because they did not know about the possibility of transforming mass into energy; for that they had to wait for Einstein.

As is often the case, the key to understanding the age of the Sun, and its power, came from another area of science. During an experiment designed to study the mysterious X-rays discovered the previous year by Wilhelm Röntgen, Henri Becquerel stored some uranium-covered plates in a desk drawer next to photographic plates wrapped in dark paper. Because it was cloudy in Paris for a couple of days, Becquerel was not able to 'energise' his photographic plates by exposing them to sunlight as he had intended. But on developing the plates, he saw to his great surprise strong images of his uranium crystals. He had discovered radioactivity.

The significance of the discovery became apparent in 1903, when Pierre Curie and his young assistant, Albert Laborde, announced that radium salts constantly release heat. The most extraordinary thing was that the radium radiated heat without cooling down to the temperature of its surroundings, revealing that it had a previously unknown source of energy keeping it warm. Almost immediately it was proposed by George Darwin, Charles Darwin's son, that this 'radioactivity' might be the source of the sun's energy. But it was Ernest Rutherford, then a professor of physics at McGill University in Montreal, later at Manchester University and then Cambridge University, who discovered the enormous energy released from radioactive substances. In 1904 he wrote:

The discovery of the radio-active elements, which in their disintegration liberate enormous amounts of energy, thus increases the possible limit of the duration of life on this planet, and allows the time claimed by the geologist and biologist for the process of evolution.

What the Sun is made of

It was radioactivity that freed theorists from relying in their calculations on gravitational energy to explain the age of the Sun and the source of its power. Mass for mass, radioactivity has a million times more energy than chemical energy and a hundred times more than gravitational energy. But this was not the final explanation, just a hint of the right direction, as subsequent observations showed that the Sun does not contain a lot of radioactive materials. Instead, it is mostly hydrogen. Something other than radioactivity was required to release nuclear energy within a star. Enter Einstein. In 1905, Albert Einstein derived his famous relation between mass and energy, $E = mc^2$, as a consequence of his special theory of relativity. The equation showed that a tiny amount of mass could, in principle, be converted into a tremendous amount of energy.

Eddington had a long-running argument with fellow English astronomer James Jeans over the mechanism by which energy was created in stars. He wrote, correctly, that as to the process of generating energy, 'probably the simplest hypothesis ... is that there may be a slow process of annihilation of matter'. Jeans, a little behind the times, still favoured the theory that the energy was the result of contraction.

In his monumental monograph in 1926, 'The Internal Constitution of the Star', Eddington described the energy and pressure balance of stars, providing mathematical models of temperature and density inside stars. From his calculations he estimated that the temperature at the centre of the Sun was 40 million degrees and he found a simple relationship between a star's mass and its luminosity. But his calculations were independent of the precise energy source. As for the source of that energy, Eddington had two ideas. One was the energy from the annihilation of electrons and protons; his second idea was the construction of heavy atoms by the fusion of protons with some conversion of mass to energy.

When Eddington offered these ideas many physicists did not think they would work, as they assumed that the Sun had the same general composition as the Earth. The argument went that the Sun and the Earth had both condensed from the same nebula, so they must have the same composition, therefore to determine the composition of the Sun all you had to do was to look at what the Earth was made of. This was wrong and the realisation that it was wrong came out of India.

Indian physicist Meghnad Saha (1894–1956) was one of the few who began the new age of astrophysics – the application of physics, developed in the laboratory, to the rest of the Universe. Born in Dacca, the son of a shopkeeper, he was lucky to go to school. His father could not afford to send him but a local businessman paid for him until he won a scholarship. Nevertheless, he was soon expelled for helping to organise a boycott of a visit by the governor of Bengal. Eventually, however, he went to Calcutta's University College of Science.

Saha was very interested in the way electrons were stripped from atoms – ionisation is the technical term. He also noticed that in the spectra of very hot stars all he could see was the lightest elements, hydrogen and helium, while cooler stars showed spectral lines from many elements. He developed an equation, called the Saha ionisation formula, that indicates how the degree of ionisation of an element depends on temperature. This provided a much-needed way to determine how much of each element there was in a star's atmosphere. It was the key to cracking the Sun's spectral code.

It was in 1925 that Cecelia Payne used Saha's equation, and improvements on it by Fowler and Milne, to analyse the solar spectrum and she concluded that it was mostly hydrogen. Payne had been born in England and went to Cambridge University. In 1919 she attended a lecture given by Eddington and decided to devote herself to astronomy. She wrote that it was 'a complete transformation of my world picture.' But Cambridge did not have an astronomy course at the time so she had to read physics. She felt oppressed by the male-dominated world that was British astronomy at the time, when women were not awarded degrees, so she moved to Harvard University in America. One of the reasons she went there, she later said, was that it housed the world's largest collection of stellar spectra.

Payne produced a doctoral thesis in 1925 entitled 'Stellar Atmospheres', which was later described as 'the most brilliant PhD thesis ever written in astronomy'. She proved that the Sun was mostly hydrogen and some helium. But brilliant or not, some were unimpressed. Henry Norris Russell (1877–1957), Princeton astronomer and one of the most influential astronomers in the world, ridiculed the idea: the discovery was, after all, made by a graduate student and a female at that! Under pressure Payne said that her results were 'spurious'.

In science authority is a double-edged sword. Albert Einstein once said that he had achieved much because of his contempt for scientific authority. Then he added that it was a cruel twist of fate that given his contempt for authority, fate had made him an authority himself. 'Whoever in discussion adduces authority uses not intellect but memory,' Leonardo da Vinci once said.

In the years following Payne's thesis, Henry Norris Russell was working on the Sun's spectrum at the Mount Wilson Observatory in California. With his assistants, one male, one female, he measured the Sun's spectral lines and determined a way to measure the abundance of the elements on the Sun's surface. The result was that it was mostly hydrogen, but he still refused to believe it.

Three years after Payne's thesis, Albrecht Unsold, also at Mount Wilson Observatory, used the new science of quantum mechanics to prove Payne was right. Incidently, Unsold's subsequent analysis of the spectrum of the B0 star Tau Scorpii, obtained on a 1939 visit to the Yerkes and McDonald Observatories, provided the first detailed analysis of a star other than the Sun. The evidence that the Sun was mostly hydrogen with some helium was now overwhelming and Russell had to admit that Payne was correct. Payne's groundbreaking work on stellar atmosphere led to her being included in the list of American Men in Science for 1926; at the age of 26 she was the youngest scientist included up to then. She married Sergei Gaposchkin and together they did outstanding work on the subject of variable stars. But for the first 13 years at Harvard she was merely a 'technical assistant' to the director. It was only in 1956 that she became a professor.

The Sun's energy

How does the Sun produce its energy? There were some clues. In 1919, Henry Norris Russell said that the most important clue was the high temperature in the interiors of stars. The following year F. W. Aston in Cambridge discovered the crucial piece of the puzzle.

Francis William Aston (1877–1945) is one of the great heroes of the Sun's story and indeed of physics in the twentieth century. A dignified, grey-haired figure in a sports jacket, he could often be seen making his way unobtrusively across the Great Square of Trinity College, Cambridge. Of the nearly 300 stable isotopes of the elements, he isolated and measured the relative masses of more than 200. Giving carbon 12 units of mass, he found that hydrogen was nearly one unit and helium nearly four units. He discovered that helium had nearly 1% less mass than the sum of four hydrogen atoms. It means that when you make a helium atom out of hydrogen atom nuclei, which are protons, you liberate energy. Aston rightfully won the Nobel Prize for Chemistry in 1922. The importance of Aston's measurements was immediately recognised by Sir Arthur Eddington as the solution to the mystery of the Sun's power.

In his 1920 presidential address to the British Association for the Advancement of Science, Eddington said that that Aston's measurement of the mass difference between hydrogen and helium meant that the Sun could shine by converting hydrogen atoms to helium. This burning of hydrogen into helium would (according to Einstein's relation between mass and energy) release about 0.7% of the mass equivalent of the energy. In principle, this could allow the Sun to shine for about 100 billion years. He then went on to say:

> If, indeed, the sub-atomic energy in the stars is being freely used to maintain their great furnaces, it seems to bring a little nearer to fulfilment our dream of controlling this latent power for the well-being of the human race – or for its suicide.

With the hindsight of the rest of the twentieth century, it was a remarkable and frightening insight. But it was not a complete explanation for the energy source of the Sun. In 1920 Eddington's colleagues at the Cavendish Laboratory in Cambridge did not think that conditions at the heart of the Sun were hot enough for nuclear fusion. His reply was that they should go and find a hotter place!

The problem was that the hydrogen nuclei or protons that have to come together to merge had the same electrical charge and so repelled each other forcefully. In 'classical' physics the probability that two positively charged particles get very close together is zero, but this was not classical physics. The new discoveries that everything came in lumps or 'quanta' changed everything. In 1928 George Gamow, the great Russian-American theoretical physicist, derived a quantum-mechanical formula that gave a nonzero probability of two charged particles overcoming their mutual repulsion and coming very close together. This quantum mechanical probability is now universally known as the 'Gamow factor'. He showed that protons could come together, merge and release a little bit of energy.

Following Gamow's work, Atkinson and Houtermans and later Gamow and Edward Teller used the Gamow factor to derive the rate at which nuclear reactions would proceed at the high temperatures believed to exist in the interiors of stars. The Gamow factor was needed in order to estimate how often two nuclei with the same sign of electrical charge would get close enough together to fuse and thereby generate energy according to Einstein's relation between excess mass and energy release. In 1938, C. F. von Weizsäcker came close to solving the problem of how some stars shine that do not use the proton–proton cycle that Eddington and Gamov described. He discovered a nuclear cycle, now known as the carbon–nitrogen–oxygen (CNO) cycle, in which hydrogen nuclei could be burned using carbon as a catalyst. It is a minor source of energy for our Sun, but is far more important for larger stars that have hotter cores.

By 1938 Hans Bethe, one of the greatest nuclear physicists ever, had completed a classic set of three papers in which he reviewed and analysed all that was then known about nuclear physics, a work known as 'Bethe's bible'. Following this Gamow assembled a small conference of physicists and astrophysicists in Washington, DC to discuss the state of knowledge, and the unsolved problems concerning the internal constitution of the stars. After the meeting Bethe calculated the basic nuclear processes by which hydrogen is fused into helium in the centre of the Sun. He described the results of his calculations in a paper entitled 'Energy Production in Stars', which is remarkable to read. The search was over. The Sun and stars shine because they have nuclear fusion reactors at their hearts.

In the first two decades after the end of World War II, many important details were added to Bethe's theory of nuclear burning in stars. Distinguished physicists and astrophysicists, especially Cameron, Hoyle, Salpeter and Schwarzschild, continued the investigation of how stars like the Sun generated energy. W. A. Fowler – Willy, as he was universally known – led a team of colleagues in his Caltech Kellogg Laboratory and physicists throughout the world to measure or calculate the most important details of the p–p chain and the CNO cycle. The work mostly done, the scientists moved to the next problem, how the heavy elements, which are needed for life, are made in stars.

The Sun even came with a self-control mechanism to prevent it exploding. It was explained in 1935 by English theorist Thomas Cowling that a tiny increase in the core temperature would lead to a huge increase in the energy output, which would lead to another temperature increase and a runaway explosion. But in reality the core gases would expand when heated and push against the overlying layers and the energy needed for this would cool the core, keeping the Sun stable. The Sun is a machine for generating energy, a machine that self-regulates so that it can remain stable for billions of years turning matter into light, and this happens in every corner of the universe.

Eddington's popular writings are remarkable and still worth reading today and he was the source of many profound and thought-provoking quotes about science. He also had a rather unusual view of the importance of the history of a subject, believing that knowing history was a hindrance to research. I know some science journalists who think the same.

The hard-headed mathematician and down-to-earth astronomer possessed a mystical side and the last years of his life were spent in an attempt to construct a synthesis of the physical Universe. From the late 1920s, his efforts were increasingly devoted to what later acquired the name of an unfinished and posthumously published book, *Fundamental Theory* (1946). He was not unlike Einstein (or for that matter Faraday, nearly 100 years earlier) in seeking a theory that would unify everything, the electromagnetic interactions, governed by quantum mechanics, and gravity, described by general relativity. But Eddington, like Einstein, failed. Einstein was well aware of his failure, while Eddington thought that he was succeeding.

He once wrote, 'I believe there are 15,747,724,136,275,002,
577,605,653,961,181,555,468,044,717,914,527,116,709,366,
231,425,076,185,631,031,296,296 protons in the universe and
the same number of electrons.'

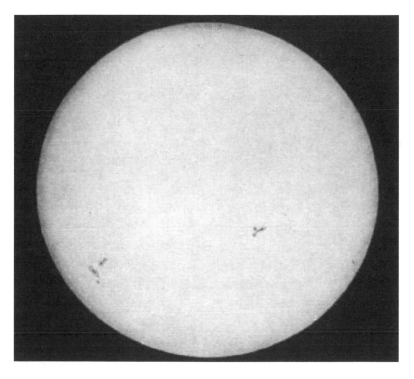

The first successful photograph of the Sun 2 April 1845.

CHAPTER SEVENTEEN

White fire laden

On the rare occasions I see the northern lights from my home in southern England, I think of a few lines from one of my favourite poets, Shelley, who in 'The Witch of Atlas' writes about a lovely lady, garmented in light from her own beauty. It is a wonderful, eerie and magical sight, those dancing and swirling mists of light you can often see at night most often from high latitudes. Veils and curtains of colour rippling across the dark sky, green and red, waving and swirling above you, and as the lights fade away the dark night closes over you once again.

In Norse mythology a bridge named Bifrost connects the Earth and Asgard, the home of gods. It was probably modelled after rainbows or perhaps the aurora. Some cultures thought that the aurora was a narrow, torchlit celestial pathway for departed souls travelling to heaven.

Birth and death are often associated with it. Some cultures, especially those at high latitudes, have regarded the sighting of the aurora as a sign of royal birth; to others it suggests ghosts of the dead. In Scandinavia the aurora was linked to dead women, especially dead virgins, but in China it was believed to predict forthcoming births. Early dragon legends of China and Europe are said to have come from the aurora. To the Iglulik Inuit, Arshät was a powerful spirit of the aurora who assisted shamans.

The earliest descriptions of the aurora come from the Mediterranean countries and from ancient China. The Old Testament says 'a whirlwind came out of the north, a great cloud, and a fire infolding itself, and a brightness was about it, and out of the midst thereof as the colour of amber, out of the midst of the fire' (Ezekiel 1:4).

In 344 BC, the Greek philosopher Aristotle observed the aurora and compared its light with flames from Earth. According to Seneca, Tiberius Caesar mistook their red glows for a fire and dispatched an army to Ostia in 34 AD to inspect the damage. In 1583, thousands of French villagers made pilgrimages to the church in Paris after seeing 'warnings from Heaven' and 'fire in the air'. But until the nineteenth century no one seriously thought that they could hurt anybody – unless, as an Old Norse legend says, you happened to whistle at one!

Explorers in the Polar Regions have left us many accounts of the aurora. Firdtjof Nansen, who tried to reach the North Pole aboard his ship *Fram* in 1895–96 but was blocked by ice, was a talented artist who left many woodcuts and paintings depicting the aurora. The southern hemisphere has its aurora australis. Captain James Cook left one of the first European accounts of these southern lights. He wrote:

> In the night we had fair weather and a clear serene sky; and between midnight and three o'clock in the morning, lights were seen in the heavens, similar to those in the northern hemisphere, known by the name of Aurora Borealis, or northern lights; but I had never heard of the Aurora Australis being seen before. The officer of the watch observed, that it sometimes broke out in spiral rays and in a circular form; then its light was very strong and its appearance beautiful. He could not perceive it had any particular direction; for it appeared, at various times, in different parts of the heavens and diffused its light throughout the whole atmosphere.

Robert Falcon Scott, an English polar explorer, wrote of the aurora while in Antarctica, 'they were bent in fantastic convolutions, some curling to spiral columns. In a few moments all this had come and gone and the broad arch of a corona seemed to be rushing toward me from the south.' Other explorers, in their race for the South Pole, left descriptions in their journals as well. Shackleton saw the aurora – like Shelley's 'orbed maiden' in 'The Cloud' he wrote in his diary about white fire laden. Juha Kinnunen of Finland lives in the realm of the polar lights and writes:

> Every aurora is an individual, and some of their beauty cannot be captured on film, but sometimes you can get an image that goes some way to showing their mystery and wonder.
>
> Noora, my daughter, and I had driven some 60 kilometres north of my home town, Jyvaskyla, which is itself some 270 km north of Helsinki. At first, it was cloudy and raining, but the Finnish Meteorological Office told me that I would eventually find a clear sky. We drove a further 10 km along minor roads through the wooded landscape. Finland is a very sparsely populated country, all lakes and trees – only very occasionally did we see a farm. From these latitudes, we get Northern Lights about 15% of the time. Travel further north, to Lappland, and you get them 75% of the time.

For Juha and Noora Kinnunen these drives into the country are a joy and part of a lifetime's fascination with the northern lights:

> It is something to do with the forces and masses involved, and the fact that the cause of these night time lights is a disturbance on the surface of the Sun. It's a great natural spectacle. Watching them over the dark forests of central Finland, away from people and light pollution, is awesome.

Some time earlier, in 1906 and not far away from Finland, there was great excitement in a laboratory in Oslo University concerning an experiment to show how the aurora was caused by the Sun. Scientists from all over the campus had congregated in the laboratory, which was full of all sorts of electrical apparatus such as sparking machines, voltage generators and cathode tubes. It would not have done them much good to probe too deeply into this laboratory. On a shelf, unprotected, was some radium donated by Marie Curie that was silently giving off a highly damaging dose. But that morning they were waiting in anticipation.

Norwegian Kristian Birkeland (1867–1917) was a remarkable, if slightly odd scientist who was fascinated by the northern lights. He organised several expeditions to Norway's high-latitude regions where he established a network of auroral observatories that also collected magnetic field data. But he had other interests as well. In 1900 he had obtained patents on what we now call an electromagnetic rail-gun that used an electromagnetic field to fire a projectile, and formed a firearms company to make and market it. The rail-gun worked but the results were disappointing: his projectiles did not fly that far. So he renamed the device an aerial torpedo and arranged a demonstration with the intention of selling the company. However, at the demonstration one of the coils in the rail-gun blew up with much noise, flame and smoke. It could easily have been repaired, but an engineer called Sam Eyde told Birkeland that there was an industrial need for the biggest flash of lightning that can be brought down to Earth in order to make artificial fertiliser using a nitrogen fixation process, which they did.

That morning Birkeland was dressed in his customary fez and red leather, long, pointed slippers, for he was an Egyptian enthusiast. He told his expectant audience that he had fashioned a cathode to be the Sun and had magnetised a metal ball to be an anode Earth, which he called a terrella. Both were placed inside a vacuum chamber and the air pumps worked hard so that the electrical current could flow unhindered between cathode and anode, or rather the Sun and the Earth.

As the air pressure in the chamber decreased the Sun-cathode started to glow. Then Birkeland switched on the electromagnet in the Earth-anode and a purple glow could be seen encircling the

Earth at the equator. As the strength of the magnetic field was increased around the miniature Earth, the equatorial circle divided and became two oval rings of glowing, phosphorescent light around the magnetic poles, looking like miniature aurora. The audience was mesmerised. What Birkeland had done was to show how cathode rays, as they were called in those days, if given off from the Sun would make their way to the Earth's magnetic poles.

However, not all scientists were convinced. Their main objection was that if only cathode rays (which were later identified as electrons) were given off by the Sun then it would, over time, become highly charged. But Birkeland realised that the Sun had to radiate both positive and negative charges at the same time, although that it was mainly the electrons that produced the aurora. He even suggested that the cathode rays that caused aurora were forced into space from the regions around sunspots and that they were also responsible for magnetic storms on Earth.

He wrote a wide-ranging book that contained chapters on magnetic storms on the Earth and their relationship to the Sun, the origin of the Sun itself, Halley's comet and the rings of Saturn. Today we call Birkeland's electrical currents circling the polar regions auroral electrojets; these became the source of a controversy that continued for a quarter of a century, because the existence of these currents from the Sun could not be confirmed from ground-based measurements alone. Proof came from satellites. A magnetometer onboard a US Navy navigation satellite, launched in 1963, observed magnetic disturbances on nearly every pass over the high-latitude regions of the Earth, as Birkeland had predicted. Norway is proud of its son and has placed him and his experiments on a 200 kroner banknote.

So it is much more than light that the Sun gives the Earth. There are other forms of radiation unseen by the eye as well as clouds of charged particles.

The billowing tails of comets that point away from the Sun are testimony to the material that streams away from it. The pressure of sunlight is far too weak to push their tails of gas and dust away.

There is a stream of particles, the so-called solar wind, that is responsible. It was German physicist Ludwig Biermann who in the middle of the twentieth century proposed that a stream of high-speed charged particles came off the Sun. The solar wind was measured by the spacecraft *Mariner 2* en route to Venus in 1962. The solar wind – mostly protons – streams off the Sun in all directions at speeds of about 400 km per second. The source of the solar wind is the Sun's hot corona. We now know that the temperature of the corona is so high that the Sun's gravity cannot hold on to it.

What all this means is that above our day-to-day weather, which is confined to the lowest levels of our atmosphere, there is another type of weather going on far above our heads. It is space weather, caused by the interaction of the solar wind on the Earth's thin upper atmosphere and magnetic shroud. Radiation from the Sun affects the space weather outlook, as does the inconstant solar wind and the sometimes spectacular arrival of plasma clouds from the Sun.

The boundary between the solar wind and the Earth is the magnetopause. It is a complicated region where the Earth's magnetic field – squashed and distorted on the side facing the solar wind – holds back the stream of particles. In 1961 *Explorer 10* was launched into a very eccentric orbit in the anti-sunward direction. Only 52 hours of data were obtained on this battery-powered mission, which provided measurements out to 43 Earth radii. Part of the orbit skimmed what we know now to be the tail of the magnetopause. While *Explorer*'s data provided little resolution on the structure of the magnetopause, they did reveal that it was constantly in motion, wafting back and forth through space faster than *Explorer* was travelling. First it was one side of the magnetopause, then the other.

In the direction facing the Sun, the magnetopause is between two and three Earth radii away, changing as the pressure from the solar wind rises and falls. In front of the magnetopause by about two Earth radii is a so-called standing shock front, like the shockwave formed ahead of a supersonic airplane. Here the solar wind is forced to slow down. There are points in the pear-shaped magnetic field around the Earth called cusps.

When spacecraft were sent to these cusp regions – the European *HEOS 1* and *HEOS 2* (in 1968 and 1972 respectively) – they detected a disordered magnetic field that astronomers believe signified a weak spot in the magnetopause that lets the solar wind in, allowing it to funnel down onto the Earth's polar regions. Some of the charged particles from the Sun travel all the way to the top of the atmosphere, converging on the polar regions. As they enter the thickening atmosphere starting at about 700 km high, they energise atoms and molecules of oxygen and nitrogen, which shed their added energy as light, producing strange moving shapes and colours. Most aurora billow about 100 km above us.

But there is more going on with aurora than meets the eye.

Joseph Von Fraunhofer wizardry with lenses enabled scientists to find out what the Sun was made of.

Arcades of magnetic loops containing hot gas rise above the Sun's surface, as observed by the *TRACE* satellite.

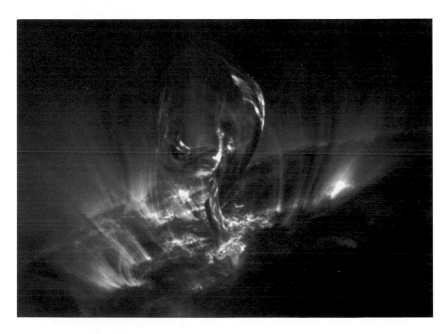

Magnetic chaos on the Sun's surface as magnetic loops emerging from below the surface constrain the hottest gases. A *TRACE* satellite observation.

Pharaoh Akhenaten and Nefertiti worship under the light of the one god Aten-Ra.

A spectacular aurora as seen by Juha and Noora Kinnunen in Finland.

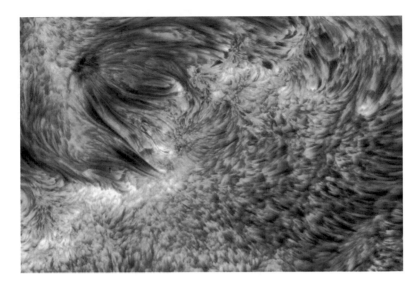

One of the most detailed images ever taken of the Sun's surface. The mysterious spicules, as wide as a country and half as long as the Earth, as seen by the Swedish Solar Telescope on La Palma.

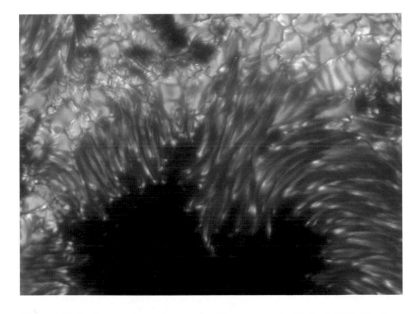

The outskirts of a sunspot where upwelling magnetic Fields inhibit the flow of outward energy, slightly cooling the Sun's surface. This spot, seen by the Swedish Solar Telescope, is larger than the Earth.

A full disk image of the Sun taken by the *Soho* satellite showing clouds of hot gas over active regions. Note the dark region, which is a coronal hole, a source of the solar wind.

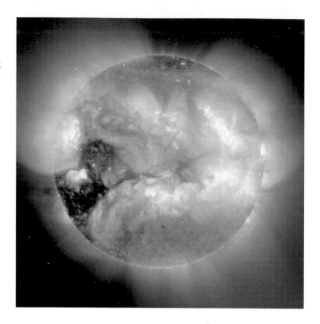

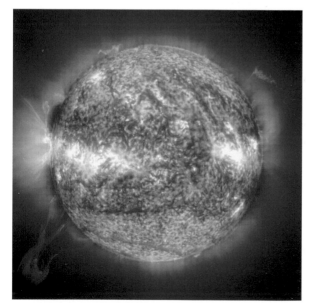

Huge prominences erupt from either side of the Sun, as observed by the ultraviolet telescope onboard the *Soho* satellite.

Almost perfect symmetry. A loop of magnetic field stretches over the Sun's surface, becoming filled with hot gas. Such loops are the site of solar flares, huge explosions on the Sun's surface caused by the collapse of magnetic energy.

Magnetic chaos on the Sun's surface as magnetic loops emerging from below the surface constrain the hottest gases. A *TRACE* satellite observation.

A colour composite image of the Sun made by superimposing images taken by the *Soho* satellite, showing swirls of hot gas above active regions on large and small scales.

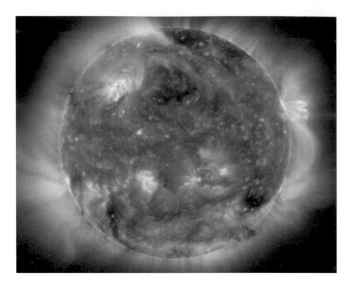

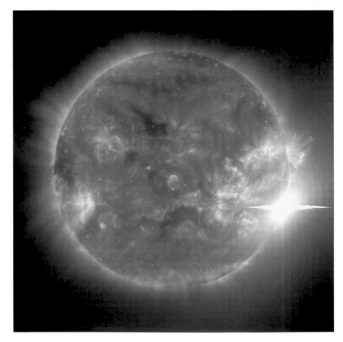

One of the largest flares ever seen to erupt on the Sun's surface, 4 November 2003, as seen by the *Soho* satellite.

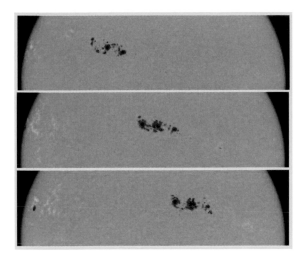

These spots cover an area nearly 10 times the size of Earth. They were seen on 15, 16 and 17 July 2002 as the Sun's rotation carried them across the disk.

Man unleashes the power of the Sun. The first H-bomb exploded at Enewetak Atoll in the Pacific, 1 November 1952.

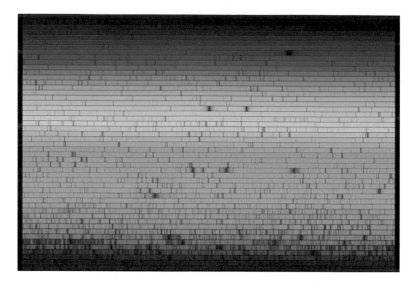

All the visible colours of the Sun's spectrum as created by the McMath-Pierce Solar Observatory. Our yellow star emits light of every colour, though it is not know why some colours are missing.

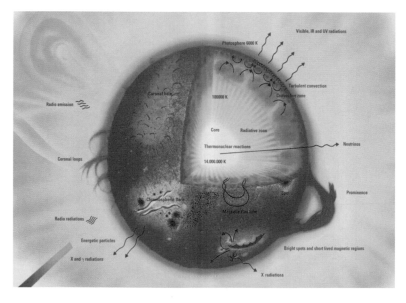

A cutaway of the Sun showing its internal structure, in particular the transition between the radiative region and the convection zone and the magnetic phenomenon on its surface. (*Scientific American*)

CHAPTER EIGHTEEN

Quebec

As Birkeland predicted, the particles from the Sun that penetrate the Earth's magnetic defences form great electrical circuits that circulate in space like a 15,000-km-wide river, acting like a generator with the current flowing down onto the Earth. It streams downwards from the magnetosphere, around the so-called auroral oval that rings the poles and back up into the magnetosphere. Electrons and protons travel in clouds, producing sheets of light when they strike atmospheric atoms – greenish blue due to oxygen, red due to nitrogen. These invisible stratospheric rivers of electricity, a million amps of current, can alter the Earth's own magnetic field temporarily and produce 'magnetic storms' where a million megawatts of power are unleashed.

In the nineteenth century English physicist Michael Faraday discovered that if you take a magnet and move it near a loop of wire, electrical current flows in the wire. Faraday's 'magnetic induction' was soon put to use in the first electric generator. This is the effect that occurs on a grand scale as the changing magnetic field around the Earth in turn induces currents in the Earth's crust as well as in good electrical conductors, such as railway tracks, pipelines, transoceanic cables and power lines.

Currents induced in the Trans-Alaska and other pipelines can promote corrosion that limits their lifetimes. On 4 June 1989, a gas pipeline explosion demolished part of the Trans-Siberian Railway, engulfing two passenger trains in flames and killing 500 people. It is thought that magnetic storm damage may have contributed to the accident. Unlike the Siberian pipeline, the

Alaskan oil pipeline built during the mid-1970s is of a newer technology specifically designed to minimise corrosive currents, now well known to modern pipeline engineers.

These currents were also a problem for the telegraph operators of the Wild West. Following the invention of telegraphy in 1841 and the telephone in the 1870s, vast systems of telephone and telegraph lines were strung up across many continents. During solar storms the currents induced by the Earth's changing magnetic field were so powerful that telegraphers didn't need a battery to send their messages down the line. Some were almost electrocuted! In the Atlantic cable between Scotland and Newfoundland, voltages up to 2600 volts were recorded during the March 1940 magnetic storm. The storm caused disruption of electrical services in New England, New York, Pennsylvania, Minnesota, Quebec and Ontario.

Power transmission lines offer a tempting target for the Sun because they are connected to the ground by the neutral wires of transformers, therefore offering a path of least resistance. The strong current induced by magnetic storms flows through transformers that were not designed to handle this type of current, causing voltage fluctuations and dropouts. It is a problem that has affected all power utilities located at high latitudes.

On 9 March 1989, astronomers at the Kitt Peak Solar Observatory in Arizona spotted a major solar flare in progress. The Earth's outer atmosphere was struck by a blast of powerful ultraviolet and X-ray radiation. The next day, an even more powerful eruption launched a cloud of gas 35 times the size of the Earth from Active Region 5395. The storm cloud rushed out from the Sun at nearly two million km an hour, and on the evening of 13 March it reached Earth.

Observers in Scandinavia and Alaska were treated to a spectacular aurora seen as far south as Florida, Cuba and Mexico. Some who had never seen such a thing before were worried that a nuclear strike might be in progress. Some wondered about the fate of the space shuttle *Discovery*, which had been just been launched.

The cloud of plasma from the Sun ploughed into the Earth's magnetic field and caused a powerful jet stream of electric current to flow 100 km above the ground. It coiled, gyrated and stretched deep into North America. Invisible electromagnetic forces were doing battle in the glowing sky and in the subterranean reaches of the Earth itself as powerful electrical currents in the ground surged through the US, UK, Scandinavia and Canada.

Quebec sits on an immense geological shield that prevents current from flowing easily into the ground, forcing the current to seek an outlet – power lines. In addition, Hydro-Québec's power grid has very long transmission lines, which increased the currents that can be induced in them.

At 2:44:17 a.m. those currents found a weak spot in the Quebec power grid. Static Volt-Ampere Reactive capacitor Number 12 at the Chibougamau substation tripped and went offline. The loss of voltage regulation at Chibougamau created power swings and a reduction of power generation in the 735,000-volt La Grande transmission network. Two seconds later a second capacitor failed at Chibougamau. Then, 150 km away, capacitors tripped at the Albanel and Nemiskau stations, followed swiftly by the Laverendrye complex to the south of Chibougamau.

In less than a minute Quebec lost half its electrical power-generation system. Automatic load-reduction systems tried to restore a balance between the loads, but it was an unequal struggle. One by one towns and regions across Quebec went dark as the entire power grid collapsed. In Toronto the night-time temperature was −6.8 degrees Centigrade. Millions were affected. Montreal's underground walkways were plunged into complete darkness. The Metro shut down, as did Dorval Airport. Traffic lights failed. It lasted nine hours and the total estimated cost was 3–6 billion dollars. But it was not just Canada that was affected.

The New York Power authorities lost power the moment Hydro-Québec went down, as did the New England Power Pool.

However by 9:00 a.m., New York Power and NEPool were sending over 1100 megawatts of power to Quebec to tide it over while the system was being brought back up again. Luckily, these states had the power to spare at the time, but only just. In fact, the electrical power pools serving the entire northeastern United States came very close to going down. Some 1500 km away Allegheny Power, which serves Maryland, Virginia and Pennsylvania, lost 10 of its 24 capacitors. Across the United States over 200 transformer and relay problems occurred within minutes of the start of the storm. Around 50 million people in the United States went about their business or slept, never suspecting that their electrical systems had been driven to the edge of disaster. All that prevented them from joining their Canadian friends in the dark were a dozen or so heroic capacitors on the Allegheny power network.

Automatic garage doors in California's suburbs began to open and close without apparent reason as offshore navy vessels switched to low-frequency backup transmitters. Microchip production in the northeastern United States came to a halt several times because the ionosphere's magnetic activity resulted in poor-quality wafers. In-orbit communication satellites that used the Earth's magnetic field to point themselves tried to flip themselves upside down. A few satellites in polar orbits tumbled out of control for several hours. Weather satellite communications were interrupted, causing weather images, used for daily forecasts, to be lost. NASA's *TDRSS-1* communication satellite recorded over 250 electrical and communications incidents.

In low-earth orbit the space shuttle *Discovery* was having its own problems. 'The hydrogen is exhibiting a pressure signature that we haven't ever seen before,' said the flight director, Granville Pennington, at the Johnson Space Center.

Space storms don't even have to take a direct swipe to harm a satellite. Atmospheric friction causes other headaches. During the Quebec blackout the US Space Command had to recompute orbits for more than 1300 objects affected by momentarily increased air resistance after the added energy puffed up the atmosphere like a balloon. The March 1989 Quebec blackout has reached legendary stature among electrical engineers and space scientists, as an example of how solar storms can affect us. Fortunately, storms as powerful as this are rare, occurring once every 10 years or so.

We have weathered half a dozen solar activity cycles since the end of World War II, but as solar maximums and solar storms come and go the world has been changing, becoming more reliant on electronics, satellites and communications. In 1981, at the peak of Solar Cycle 21, there were 15 communication satellites in orbit. Mobile phones were rare and there were less than a million personal computers sold in the US and only 300 hosts on the Internet. By the time the peak of the next solar cycle came around in 1989, there were 102 communication satellites and 3 million cellular phone users in the United States alone, as well as 300,000 host machines on the Internet.

The world had changed even more as we passed through the peak of the 23rd Sunspot Cycle in 2000–2001. Now there were 349 communication satellites. Hundreds of millions of people have cellular phones, and Global Positioning System handsets are routinely used for work and leisure. There are over 10 million Internet hosts. More electricity than ever is flowing in the world's power grids.

In May 1998 when the *Galaxy IV* satellite tumbled out of control for a while, over 90 million pagers in the US were silenced as well as radio and television feeds and newspaper services. Just one satellite failure did this. The US electric system includes over 6000 generating units, more than 800,000 kilometres of bulk transmission lines and approximately 12,000 major substations, all served by 100 separate control centres. All are vulnerable to solar attack. We have learned to cope with much of what the Sun has thrown at us, but there must be more energetic storms that occur every 100 or 1000 years on their way that will surprise us.

The sound of aurora

There is a curious myth that sometimes aurora can make sounds. You can hear the sounds of aurora with a radio receiver. As electrons spiral through the Earth's magnetic field they give off radio waves that change in frequency, called 'whistlers.' These screeches, descending in pitch alongside strange static sounds, feel like having an ear to another world. Research teams string out antenna to capture the sounds and many enthusiasts travel the polar wastes in search of them.

Some say that, when conditions are right, you do not need a radio receiver to hear the aurora. Reports going back centuries say that there are two types of sounds reported to accompany the aurora. The first is a swishing sound that changes with movements in that auroral display. The second is a crackling sound.

But the aurora is a long way away – 100 km above our heads – so if sound was coming from it there would be a long delay between the auroral movement and the sound arriving, just as thunder arrives long after a distant lightning flash. Also the air between the listener and the aurora is far too thin to carry sound over such long distances. Therefore, if this kind of sound exists, it must be created very near to the observer. It has been suggested that it is possibly inside the observer's head, caused by the leakage of the electrical impulses from the nerves in the eye (carrying images of the aurora to the brain) into the part of the brain where the processes sound. In a very quiet environment there are no sound signals for the brain to process, so it notices these tiny leakage signals and the result is sounds that change in time with the aurora. This explanation was actually tested by some early explorers, who found that the sound went away if their eyes were covered.

In 'Jane, the keen stars were twinkling', Shelley writes of a voice revealing a tone of some world far from ours, where music and moonlight combine with feeling: the strange world of the aurora perhaps.

From 1826 Heinrich Schwabe scanned the Sun almost every day for 17 years and discovered something amazing.

CHAPTER NINETEEN

Coalbrookdale by Night

The Cerdagne plateau in the Pyrenees is guarded by impressive 3000-metre peaks – the Carlit, Peric and Puigmal d'Err massifs. For centuries the border between France and Spain ebbed and flowed across this landscape of craggy peaks and forests of Scots pine, orchids and chamois and the unique culture of Catalonia. The Romans took advantage of thermal springs in the region and Louis XIV, the Sun King, fortified its borders. At Odeillo in the Cerdagne the sun shines more brightly than anywhere else on earth.

Watching the solar furnace at Odeillo swing into action is an awesome sight and a demonstration of the power of sunlight. Sixty-three mirrors – heliostats – are computer controlled to align themselves with the Sun, track it and reflect its light onto a giant parabolic wall of mirrors, which gather the energy to a focus. When the doors to the focus chamber open the heliostats are off beam, but as they find the Sun the intensity of the focused light grows to a crescendo and almost nothing can withstand it. The material of rocket nosecones is tested here, as were the tiles to protect the purported European space shuttle called *Hermes* from the fiery heat of re-entry into the earth's atmosphere. Temperatures in excess of 3000 degrees Celsius can be reached at the focus, allowing research into metallurgy and the development of heat-resistant and fireproof material. Energy can be collected from the Sun and put to use – but as usual, nature got there first.

Photosynthesis

Leaves are a wonder. They hold a special place in our lives. Many equate them with tranquillity and when asked to describe a peaceful place will say a leafy glade. The view from my study at home is mostly leafy and I feel comfortable and at ease, whereas my office in London once had a view of other offices and I hated it.

We all enjoy the show of colours as leaves change each autumn. They do so because of the Sun, because of the Earth's orbit around the Sun and because of the spectrum of the light it gives us. The Sun is the giver of life to the Earth and life feeds off the Sun by consuming its light. Leaves are one of nature's ways of harvesting energy from the Sun.

Energy is neither created nor destroyed, just moved from one form to another. What began in the Big Bang, when energy was compressed into subatomic particles, continued in the gravitational attraction of collapsing gas and dust clouds, was released in the nuclear reactions at the Sun's core, travelled for a million years to reach the Sun's surface and flashed across the inner solar system for eight minutes to the Earth, finally to fall as a beam of sunlight on a leaf where the energy begins new adventures, this time along complex biological avenues to provide the power for every living thing on Earth. All the food we eat and all the fossil fuel we burn, as coal or as oil, is a product of photosynthesis occurring recently or in ancient times.

As well as sunlight from above, plants take water from the ground through their roots. They take in carbon dioxide from the air using the energy of sunlight to turn water and carbon dioxide into glucose, a kind of sugar used as a source of chemical energy. This is photosynthesis – 'putting together with light' – which involves a complex and remarkable chemical called chlorophyll and it is no accident that it is green.

Leaves are green because they are tuned to the radiation from the Sun, absorbing energy from the region of the spectrum where it is most abundant. Photosynthesis occurs in cells or organelles – structures inside cells – that are typically only a few thousands of a millimetre across. Light energy is absorbed by pigments, primarily chlorophylls and carotenoids. Chlorophyll absorbs blue and red light and carotenoids absorb blue-green light, but green and yellow light are not effectively absorbed by the photosynthetic pigments in plants; therefore, this colour is either reflected by leaves or passes through them. This is why plants are green. For stars unlike our own type G2 yellow dwarf, the peak in the spectrum will be at other regions of the visible spectrum and their version of chlorophyll will be a different colour; if they have them, their trees and forests will be riots of red or blue.

In a year the solar radiation striking the earth is equivalent to 178,000 terawatts. The energy scooped up by photosynthesis is estimated to be 10 times more than mankind uses. Individually plants do not convert very much sunlight into energy. One of the most efficient crop plants is sugar cane, which has been shown to store up to 1% of the incident visible radiation over a year. Most crops are less productive. The annual conversion efficiency of corn, wheat, rice, potatoes and soybeans typically ranges from 0.1% to 0.4%. But because of the sheer numbers of plants put together, it is photosynthesis that keeps us and almost every living thing on Earth alive. Not only in producing food – all flesh is grass, they say – but also oxygen. An average hectare of corn produces enough oxygen per hectare per day in mid-summer to meet the respiratory needs of about 325 people.

The light from the Sun and the rhythm of the Earth's orbit drive many natural cycles. During winter there is not enough light or water for photosynthesis and then the trees will rest, and live off the food they stored during the summer. As they begin to shut down their food-making factories chlorophyll disappears from the leaves and, as the bright green fades away, we see yellow and orange undercolours painted by the diminishing sun. The bright reds and purples we see in leaves are made mostly in the autumn. In some trees, such as maples, glucose is trapped in the leaves after photosynthesis stops. Sunlight and the cool nights of autumn turn this glucose into a red colour. The brown colour of trees like oaks is made from wastes left in the leaves.

That leaves feed off sunlight is obvious, but discovering the way they do it took centuries to unravel.

In the 1770s Joseph Priestley, an English chemist and clergyman, performed experiments showing that plants release a gas that allows combustion. He demonstrated this by burning a candle in a closed vessel until the flame went out. He placed a sprig of mint in the chamber and after several days showed that the candle could burn again. Although Priestley did not know about molecular oxygen, his work showed that plants release oxygen into the atmosphere.

It was not until 1845 that Julius Robert von Mayer (1814–78), a German physician and physicist, proposed that photosynthetic organisms convert light energy into chemical energy. But what should have been his credit was given to someone else. Although he was the first to describe the vital chemical process now referred to as oxidation as the primary source of energy for any living creature, he was not taken seriously at the time and credit was given to James Joule, an English physicist, son of a wealthy brewery owner (yet another one) and pioneer in the study of energy. The fundamental unit of energy is named after him. Despondent, Mayer almost committed suicide and he spent some time in mental institutions to recover from this and the loss of some of his children. His overlooked work was revived in 1862 by fellow physicist John Tyndall in a lecture at the Royal Institution in London. In 1867, Mayer was awarded personal nobility (von Mayer), which is the German equivalent of a British knighthood. It was just in time: he died from tuberculosis a few months later.

By the middle of the nineteenth century the key features of plant photosynthesis were known: namely, that plants could use light energy to make carbohydrates from carbon dioxide and water. Most of the molecules required for the conversion of light energy in photosynthesis are located in membranes inside cells, but despite decades of work, efforts to determine the structure of the membrane-bound proteins had little success until relatively recently. This changed in the 1980s when scientists determined the structure of the so-called reaction centre of the purple bacterium *Rhodopseudomonas viridis*, a work that was rewarded by the Nobel Prize for Chemistry in 1988 for 'the determination of the three-dimensional structure of a photosynthetic reaction centre', as the citation read. Purple bacteria are great masters of harvesting light.

But how did it start? How did the fantastically complex molecule that is chlorophyll and the series of chemical reactions that is photosynthesis come about? The answer lies with the primitive Earth and the primitive Sun.

The beginnings of life

Studies of the earliest rocks show that as soon as the Earth cooled life started. How it started we do not know, neither do we know if it involved chemicals that were present on the Earth or had been deposited there by comets, but we do know that the Sun was involved. As we have seen it was then a brighter Sun, shining down on an Earth with an atmosphere of nitrogen, carbon dioxide, water vapour, methane and ammonia, and by four billion years ago it had cooled down enough for oceans to form from water accumulated from outgases or deposited here by comets.

Stromatolites are the oldest known fossils, dating back more than three billion years. They are formed by colonies of photosynthesising cyannobacteria and other microbes. Stromatolites are prokaryotes (primitive organisms lacking a cellular nucleus) that thrived in warm aquatic environments and built reefs much the same way as coral does today. Cyannobacteria were the dominant life form on Earth for over two billion years and are probably responsible for the creation of Earth's oxygen atmosphere. Today they are nearly extinct, living a precarious existence in only a few localities worldwide such as Shark Bay in Western Australia.

We now know that all life can be divided into three domains: Eucarya, which includes animals, plants and fungi among others; Bacteria; and the relatively recent classification Archaea, life's extremists that on the face of it do not look very different from Bacteria, although Bacteria and Archaea are as different from each other as they both are from Eucarya.

It seems that no one of these groups is ancestral to the others: all originated from a common ancestor billions of years ago. The seemingly most primitive domain, Archaea, includes organisms known as halobacteria that convert light energy into chemical energy. However, the mechanism by which halobacteria convert light is fundamentally different from that of higher organisms. The light-sensitive pigment bacteriorhodopsin gives some Archaeans their colour and provides them with chemical energy. Bacteriorhodopsin has a lovely purple colour and it pumps protons to the outside of the membrane. When these protons flow back, they are used in the synthesis of a chemical called ATP, which is the energy source of the cell. This protein is chemically very similar to the light-detecting pigment rhodopsin, found in the retina of our eyes. Consequently, some biologists do not consider halobacteria as photosynthetic.

However, they do give some indication of how the first primitive organisms used the light of the Sun, and as they evolved into more complex creatures the process of converting light into energy became more complicated and better able to serve a more sophisticated organism. Some photosynthetic bacteria can use light energy to extract electrons from molecules other than water, suggesting that these organisms are of ancient origin, presumed to have evolved before photosynthesis turned the Earth's atmosphere oxygen rich.

A key event in the development of life may have come about because of two separate free-living organisms floating in some ancient sea. They evolved separately but came together for their mutual advantage. The chloroplasts provide the energy and reduced carbon needed for plant growth, while the plant provides the chloroplast with carbon dioxide, water, nitrogen, organic molecules and minerals necessary for the chloroplast to survive. Organisms living near seafloor vents were once thought to be cut off from the Sun, but we now know that they are not and that they rely indirectly on the oxygen supplied by photosynthesis. Not every living thing on earth is part of a food web that relies on the light from the sun. Scientists have recently discovered microbes living inside extremely hard rocks that lack organic nutrients. In the absence of sunlight, the industrious residents of these rocks fashion their own organic molecules out of the barest of inorganic materials. Relying only on hydrogen, water and carbon dioxide – all of which are products of Earth's interior – these microbes are unique among the vast array of living species. All other organisms depend to some extent on the Sun's energy, which is harnessed through photosynthesis and creates food for surface life. Are they microbes that evolved there or moved there from elsewhere? We do not know.

For the vast majority, and possibly for all creatures, the Sun is the source of life on Earth, providing the first input into the complex food web. It was the energy of the Sun that was harnessed by the molecules of ancient Earth to organise themselves into primitive life, and it was through these ways to exploit the energy given to the Earth by the Sun that life evolved. In a way the ancient myths and legends are right: we are all creatures of the Sun.

Solar energy

Humans have tried hard to make artificial leaves. The development of the solar cell – a device that turns sunlight into electricity – stems from the work of the French physicist Antoine-César Becquerel (the discoverer of radioactivity in uranium) in 1839. He discovered the photovoltaic effect when he noticed that a voltage was produced when light fell on certain substances. About 50 years later, Charles Fritts made the first true solar cells using the metals selenium and gold. But Fritts's devices were very inefficient, transforming less than 1% of the absorbed light into electrical energy.

By 1927 another solar cell design was tried out, made of copper and copper oxide, and by the 1930s both the selenium cell and the copper oxide cell were being employed in light-sensitive devices, such as photometers, for use in photography, although they were still poor performers. This changed with the development of the silicon solar cell by Russell Ohl in 1941. To my mind Russell Ohl is one of the forgotten heros of twentieth-century physics. His discovery of the effect in silicon that produced electricity from light was crucial to the development of the transistor, a device that perhaps more than any other has changed civilisation recently. In 1954 three other American researchers, G. L. Pearson, Daryl Chapin and Calvin Fuller, demonstrated a silicon solar cell capable of a 6% energy-conversion efficiency when used in direct sunlight. But despite this encouraging trend, progress in solar cells was slow for decades, perhaps because there was no need for them, oil being so plentiful.

By the late 1980s silicon cells had been produced, as well as those made of gallium arsenide with efficiencies of more than 20%. In 1989 a concentrator solar cell, a device in which sunlight is concentrated onto the cell surface by means of lenses, achieved an efficiency of 37% and nowadays solar cells are widely available.

Today solar energy accounts for less than 0.1% of total global primary energy demand. However solar energy demand has grown at about 25% per annum over the past 15 years and of that global demand for solar photovoltaics, over 35% is accounted for by Japan, 25% by European countries and less than 15% by the USA. Japan overtook Switzerland in 2001 in terms of the proportion of solar cells installed per person in the country.

Undoubtedly we could do a lot more with solar cells, our artificial leaves. Take Freiberg's example. If there was to be a solar capital of Europe it would be Freiburg in Germany, which probably has more solar energy projects than anywhere else in Europe. It houses the solar-heated headquarters of the

International Solar Energy Society and in the streets surrounding it there are solar-powered parking meters.

Coal and oil

In the Science Museum in London there is a remarkable painting by Philip de Loutherbourg (1740–1812) that is traditionally called *Coalbrookdale by Night*. It shows smoke and flames interrupting the night sky above an iron-smelting furnace at Coalbrookdale, in the English Midlands, the home of the Industrial Revolution. It is an almost terrifying image, a rebellion against the pastoral romantic art of the time. The town is being engulfed in red flame at the start of the age of mechanisation and the industrialisation of life. The few figures in it seem dwarfed and powerless against the redness and in the foreground there is a horse-drawn cart, the horse downcast, trudging away. A trip to Coalbrookdale and the nearby Ironbridge and a walk over the world's first cast iron bridge invokes many things for me, not least because I was born and raised in nearby Birmingham among the industrialisation that this place made possible.

The industrial revolution changed the productive capacity of England, Europe and the United States. But it involved more than simply new machines and smoke-belching factories. It transformed western society and, like the Reformation or the French Revolution, no one was left untouched by the idea of progress, forever forward, forever upward. Karl Marx understood what was going on, 'Philosophers have only interpreted the world in various ways,' he wrote. 'The point, however, is to change it.' And what changed the world was energy from the Sun that we can dig up and burn – coal.

Coal is a combustible, black, sedimentary rock composed mainly of carbon formed out of plant matter that accumulated at the bottom of swamps millions of years ago, during the Carboniferous Period. At that time, the earth's climate was very favourable for plant growth. Organic matter accumulated in stagnant swamps, which were low in oxygen and thus inhibited decomposition. Eventually, seas rose or the land fell and the swamps submerged, allowing sand and other debris to be heaped over the organic material. Over hundreds of thousands of years, the organic material was compressed and transformed into coal. Coal is found beneath layers of sandstone, limestone and shale all over the world. The United States, the former Soviet Union, China and India have the largest reserves of coal, but large deposits also exist in South Africa, Australia, Germany and Eastern Europe.

At the start of the nineteenth century homes were lit by gas or by candles, wood or coal provided heat for cooking and warmth, and the horse, the bicycle and the steam train were the chief means of transport. Then another gift from the Sun was discovered – oil.

Just like coal, oil and natural gas are stored sunlight, the result of the decomposition of plants and animals deposited in shallow seas tens and hundreds of millions of years ago. When we light our gas fires or start our cars, we are using energy that has its origin at the Sun's core and that fell as sunlight onto our planet long before we evolved.

Today civilization depends on an abundant and relatively cheap supply of this second-hand sunlight. In a way the twentieth century was the century of oil. The twenty-first century will not be. Oil production is peaking; if it has not already that peak must be only a few years away and as oil production declines it could intensify a scramble by some to tie up existing reserves, leading to scarcity and higher prices, possibly recession.

Various 'experts' have been predicting the end of the oil age for more than a century and even now, no one really knows how much oil is left in the ground. Estimates in this field are really informed guesses of not only future oil finds but future world economic output and oil consumption – highly imprecise figures. Even calculating current reserves is tricky. The Royal Dutch/Shell Group, one of the world's largest oil producers, shocked the financial community when it said it had over-estimated its reserves by 20%.

The United States Geological Survey puts yearly world consumption of oil today at about 30 billion barrels. That comes out of known or proven world reserves of about 1.1 trillion barrels. By adding Canada's oil sands it could rise to 1.266 trillion. Canada is the second-largest holder of reserves after Saudi Arabia. These sands are already being exploited, but they require the injection of hydrogen to make their tar oil light enough to flow in a pipe.

It has often been said that the discovery of new reserves outpaces the growth in demand, keeping us on the right side of the consumption–resources curve. But that cannot carry on forever. In 2002 the world used four times as much oil as was newly found and the rate of discovery of worldwide oil reserves, after declining for 40 years, slowed to a trickle. There were 16 large discoveries of oil in 2000, eight in 2001, three in 2002 and none in 2003. There may be nothing major left to find. Many believe that all the giant fields, such as those in the Middle East, have already been discovered. The last major oil field, Cantarell off Mexico, was discovered in 1976.

In 2003 total world oil production reached 68 million barrels per day, not much above the 66.7 million barrels per day in 2001. Production has peaked for more than 50 oil-producing nations, including the US and Britain. China, second to the US in the consumption of oil, was a net exporter of oil until five years ago. The US Department of Energy predicts that world demand will reach 119 million barrels per day in 2025, with huge increases in China, India and other developing nations. Recently the UK government approved the development of the biggest oil deposit discovered on British territory for at least 10 years. Spoken of as a 'huge' find, it shows how our expectations have slowly changed: this 'huge' new field will supply the world with oil for five and a quarter days.

The world has used up about 930 billion barrels of oil since the 1800s. It is a finite resource. Even if the optimists are correct, we will be scraping the bottom of the oil barrel within the lifetimes of most of those who are middle-aged today. As the price of oil rises, transport and farming will be forced to contract. Our city suburbs are impossible to service without cars. High oil prices will mean high food prices and unemployment. The last five recessions in the US were all preceded by a rise in oil prices.

So it is clear that we need another source of energy if we are to continue anything like the lifestyle we have. In the short term, some contribution will be made by hydro, solar, wind, wave and geothermal systems. But even with the most optimistic estimates, they will never be able to shoulder the burden of our population's large-scale energy needs. They have a power density limit that can never be overcome and are thought to be unlikely to exceed 20% of the total energy budget by 2020. Conventional nuclear power – nuclear fission – could provide the major proportion of the future's energy needs, but it is politically unpopular.

It is obvious that mankind cannot go on like this and needs a new source of energy, if there is one, that does not have the environmental, economic and political drawbacks of oil. There is only one choice. We will have to create a piece of the Sun on Earth: harness the very power that makes the Sun shine.

Only one technology could provide the energy the world needs without the risks of global warming or the political difficulties of fission. Nuclear fusion offers the possibility of high power density, no high-level radioactive waste and no greenhouse gases. Since the birth of the atomic age it has been a holy grail. It may now be within our grasp.

'It was the first time I saw the sun rise in the West,' recalls Lemyo Enob. In 1954, when he was a boy in a South Pacific paradise, he witnessed the creation of a piece of the Sun on the Earth. 'At first, I did not know what it was, but then I understand it was a big bomb.' US comedian Bob Hope put it another way, 'As soon as the war ended, we located the one spot on earth that hadn't been touched by the war and blew it to hell.'

Today the Republic of the Marshall Islands is a nation of more than 50,000 living on 29 coral atolls and five small low-lying islands midway between Hawaii and Australia. Its remarkable beauty and culture attract tourists from around the world. But look at old maps of the region and you will soon notice that something strange has happened there. Three islands, Bokonijien, Aerokojlol and Nam, are missing and another, Bikini, is not the paradise it once was. To the native islanders it is a scorched wasteland.

On 1 March 1954, the United States conducted a nuclear test code-named Bravo. A 15-megaton hydrogen bomb detonated on Bikini Atoll, resulting in an intense fireball followed by a 35 km high mushroom cloud. A fleet of ships were engulfed in the nuclear explosion and sank to the bottom of the lagoon. Radioactive fallout spread quickly, drifting towards inhabited islands. The 'sun' that 14-year old Lemyo Enob saw that day was a thousand times more powerful than the nuclear fission blast at Hiroshima, the Japanese city where the first atomic bomb was dropped.

When it became known in 1949 that the Soviet Union had developed its own atom bomb, a few scientists, including Edward Teller and military figures in the United States, urged the US government to study the feasibility of producing a 'superbomb' – hydrogen bomb, a bomb based on nuclear fusion rather than nuclear fission. They were opposed by a group of other scientists, including J. Robert Oppenheimer. But in February 1950 President Truman, spurred by the growing Cold War, ordered a programme to build the hydrogen bomb.

Edward Teller became known as the father of the H-bomb, yet he is also remembered for his role in destroying the career of his one-time boss, Robert Oppenheimer, whose opposition of the superbomb he resented. It was Italian-born physicist Enrico Fermi who first got Teller thinking about an H-bomb. In September 1941, before the United States had even built an A-bomb, he suggested to Teller that an atomic bomb might heat a mass of deuterium (an isotope of hydrogen) sufficiently to ignite a thermonuclear reaction. In the summer of 1942, Teller joined a group of physicists who were working on the design for the atomic bomb, but he was really interested in a superbomb. He travelled to California with his old friend Hans Bethe, who later recalled:

> Teller told me that the fission bomb was all well and good and, essentially, was now a sure thing. He said that what we really should think about was the possibility of ... the hydrogen bomb.

After more than a year of work, scientists solved the technical problems involved and scheduled a test of a prototype hydrogen bomb at Eniwetok Atoll. Operation Ivy, the first ever series of hydrogen bomb tests, was carried out by Joint Task Force 132 at Enewetak Atoll in the Pacific Proving Grounds. A film made at the time shows the world's first hydrogen bomb, 'Mike', arriving at Enewetak Atoll in crates. The narrator of the film describes it as 'a thing dismembered, a sleeping giant. A robot taken apart for the long trip ... the days of preparation fly in kaleidoscopic fashion across your mind. Days now in the past. Just as Mike itself will soon be in the past.' The narrator continues:

> This is the significance of the moment. This is the first full-scale test of a hydrogen device. If the reaction goes, we're in the thermonuclear era. For the sake of all of us, and for the sake of our own country, I know that you will join me in wishing this expedition well.

The projected yield of the Mike bomb was 4 to 10 megatons, but the actual yield came in at 10.4 megatons, roughly 750 times as powerful as the atomic bomb dropped on Hiroshima in 1945. As detonation hour approaches, the narrator tells us, 'We'll soon see the largest explosion ever set off on the face of the earth.'

The second Operation Ivy test was the 'King' shot, a thermonuclear bomb air dropped on 19 November 1952. The detonation left a crater 1.5 km wide. This single blast yielded more force than all the bombs dropped by all the Allied Forces during World War II. The power of the Sun had been unleashed on Earth by the United States, and within a year the Soviets would yield it as well. In the USSR the 'Installation' was a secret city in the central Volga region. Here a special design bureau was also intent on creating nuclear weapons that harnessed the energy of the Sun. Even the name of the city was secret (40 years later it was revealed as Sarov). Soviet scientists Igor Tamm and Andrei Sakharov moved to the Installation in 1950 and while the theoretical group headed by Yakov Zeldovich continued to work on their designs for a fusion bomb, Tamm's team worked on Sakharov's own design that became the first Soviet H-bomb, successfully tested on 12 August 1953.

After the thermonuclear test of 1953, Tamm returned to Moscow to resume academic work. Sakharov stayed at the Installation, working with colleagues such as Zeldovich and David Frank-Kamenetskii. He made key contributions to the Soviet Union's first fully fledged H-bomb, tested in 1955, and to the 50-megaton so-called Czar-Bomb of 1961, the most powerful device ever exploded on Earth.

The hydrogen bomb is a thermonuclear weapon capable of devastating 150 square miles with its blast, and with its searing heat effects and radioactive fallout its destruction reaches more than 800 square miles, depending on the size of the weapon. But these terrifying weapons are puny compared to the Sun's power. The hydrogen bombs we have built release just an almost infinitesimal fraction of our star's potential. Solar flares and coronal mass ejections are currently the biggest explosions in our solar system, approaching the power in one billion hydrogen bombs. The power unleashed at the centre of our Sun is equivalent to 80 billion hydrogen bombs exploding every second. You would need to detonate a hydrogen bomb every second for the next 3200 years to equal the Sun's energy output for just one second. But what if we could tame that energy?

On 24 March 1951 the Argentine dictator Juan Perón made an astounding announcement. He said that Argentina had built an atomic power station on the Isla Heumul in Lake Nahuel Huapi that produced a new form of nuclear power – the kind that powers the Sun. The *New York Times* carried a front-page article, 'Perón announces new way to make atom yield power. Method linked to the Sun's'. The Argentine work was led by Austrian physicist Ronald Richter. Foreign scientists were highly sceptical. It seems that in 1948 Richter had offered Perón a scheme to achieve controlled nuclear fusion and obtain an inexhaustible source of inexpensive energy. Perón, naively, was willing to believe that any project undertaken by a German scientist would be successful. Besides, he had fallen out with many mainstream Argentinian scientists. So he gave Richter a blank cheque and appointed him as his personal representative in the Bariloche area. The young Richter spent $300 million (mid-1950s value) on his 'controlled fusion' project. But a matter of months after Perón's announcement Richter was jailed for misleading the president. His work was shrouded in absolute secrecy and was never published. Perón should have been more careful. In the late 1940s Richter proposed a thesis, at the German University of Prague, to detect 'delta rays' emitted from Earth!

After the fiasco Perón appointed a technical committee of five, including José Balseiro, a former faculty member at the La Plata Institute of Physics, to report directly to him on whether the Richter project should be discontinued. The group worked very hard at the Huemul facilities to reproduce the results that Richter claimed, but could not. Soon after, the Argentine government discontinued the project. The Richter affair caused considerable damage to the science and engineering sectors of Argentina's higher education system.

If you go walking on Isla Heumul you can still see the remains of Richter's twin laboratories, his chemistry lab, Richter's lab, the reactor building, Richter's house and the square. But it was not the only time that there had been a false start for fusion.

On 25 January 1958 British newspapers heralded a fusion breakthrough. 'The mighty Zeta', said the *Daily Mail*, 'Limitless fuel for millions of years.' The *Daily Herald* said, 'Britain's H-men make a sun'.

The day before there had been a press conference given by Sir John Cockcroft, the director of the Harwell laboratories. Hundreds of reporters turned up to hear what the 60-year-old Nobel Prize winner had to say. He announced that the previously secret Zeta project had produced plasma temperatures of five million degrees and held that temperature for three thousandths of a second. He added that it was his firm belief – 90% sure – that fusion reactions were taking place, a world first. On BBC television that day he said, 'To Britain this discovery is greater than the Russian Sputnik.'

However, a more sombre news release was issued later that spring. The neutrons produced by Zeta were not coming from fusion reactions. Zeta had not attained fusion. But despite the embarrassing setback, there was a spirit of optimism. In 1956, one of the United States' fusion pioneers Richard F. Post was working at the University of California Radiation Laboratory at Livermore. He wrote:

It is the firm belief of many of the physicists actively engaged in controlled fusion research in this country that all of the scientific and technological problems of controlled nuclear fusion will be mastered – perhaps in the next few years.

To make nuclear fusion work requires a re-creation of the conditions found at the centre of the Sun: that gas from a combination of isotopes of hydrogen – deuterium and tritium – is heated to very high temperatures of 100 million degrees Celsius and confined for at least one second. The centre of the Sun is at the milder temperature of 15 million degrees, but it is also denser. To achieve fusion in a thin gas requires temperatures several times that of the centre of our star. Put one way it sounds easy, but it isn't. Plasma – a superheated, ionised gas – is the slipperiest, most awkward, difficult to predict and control substance that man has ever encountered and we have still not mastered it. One way to constrain the fickle plasma is to use magnetic confinement and the most promising configuration at present is the *tokamak*, a

Russian word for a torus-shaped magnetic chamber, a design that goes back to Sakharov and Tamm.

Despite the fusion bomb programmes, in 1958 fusion research was declassified at a historic meeting of scientists in Geneva. Fusion researchers from East and West revealed their hitherto secret work. To date the largest effort to develop fusion energy has come from Jet, the Joint European Torus project that began in 1978. It is housed just south of Oxford. I have visited Jet many times and I never fail to be impressed. One feels as if one is in a cathedral dedicated to science. The Jet machine is a large tokamak device of approximately 15 metres in diameter and 12 metres high. At the heart of the machine there is a toroidal (ring-shaped) vacuum vessel of major radius 2.96 metres with a D-shaped cross-section 2.5 metres by 4.2 metres. The complex system of magnetic fields prevents the plasma from touching the walls of the vacuum vessel, as such contact would quench the plasma and stop the reactions. The main component of the magnetic field, the so-called toroidal field, is provided by 32 D-shaped coils surrounding the vacuum vessel. This field, combined with that produced by the current flowing in the plasma, the poloidal field, form the basic magnetic fields for the tokamak magnetic confinement system. Additional coils positioned around the outside of the mechanical shell are used to shape and position the plasma.

During most operations of the machine a small quantity of hydrogen or deuterium gas is introduced into the vacuum chamber and it is heated by passing a very large current (up to seven million amps for a pulse time of up to 30 seconds) through the gas, thereby creating a high-temperature plasma. The current is produced using a massive eight-limbed transformer. A set of coils around the centre limb of the transformer core forms the primary winding and the ring of plasma is the secondary. Recently currents up to 3 mega-amps have been produced by a different method using radio frequency waves, which have the advantage of not being confined to pulsed operation, as is the transformer action.

Using these heating methods plasma temperatures of 40–50 million degrees Celsius have been routinely achieved during the first phase of operation. Progressive amounts of additional heating have been provided in subsequent phases of Jet operation by injecting beams of energetic hydrogen or deuterium atoms into the plasma and by the use of high-power radio frequency waves, allowing plasma temperatures in excess of 300 million degrees Celsius to be achieved.

Jet started operating in 1983 and became the first fusion facility in the world to achieve a significant production of controlled fusion power (nearly 2 MW) with a deuterium–tritium experiment in 1991. During 1997 the Jet operations included a three-month campaign of highly successful experiments using a range of deuterium–tritium fuel mixtures. The results were of major significance, allowing Jet to set three new world records: 22 MJ of fusion energy in one pulse, 16 MW of peak fusion power and a 65% ratio of fusion power produced to total input power.

But despite its success, Jet is not big enough to be a commercial reactor. A larger reactor along the same design must be built. As I write, the world's fusion researchers will shortly decide where to build the next step in controlled thermonuclear fusion – a bigger reactor than Jet, one that can emit more energy than it consumes and can produce a self-sustaining reaction. If these objectives were achieved, the experimental basis of fusion power would be established. Fusion would no longer be a dream, it would be a reality.

In 1998 an international group of fusion scientists completed the design of the International Tokamak Experimental Reactor, called Iter, but at $6 billion, the cost frightened politicians. It scared US politicians so much that they pulled out of the project, despite the fact that it was Reagan and Gorbachev who first backed work on the reactor's development. Years later the US came back into the project.

In Iter, the energy from the plasma that is converted to useful heat energy is that released in the neutrons. The neutrons are absorbed by the surrounding wall. The material challenges of constructing this wall are considerable. The first 2 cm of the structure have to withstand high neutron fluxes, immense heat and the impact of high-energy particles. Work is currently focusing on stainless steel coated with beryllium, graphite or tungsten, but in an operating reactor the stainless steel would need to be replaced with another material, possibly silicon carbide, as steel becomes irradiated when exposed to neutrons. Beyond the first 2 cm is a 40 cm thick wall of steel cooled with water, which gets hot. This is where the neutron energy would be converted to heat to generate electricity.

Its advocates say that there are many advantages to fusion power. First, only a small quantity of fuel is injected into the reactor at any one time, so if something goes wrong the fusion reaction is quickly quenched. Secondly, the fuel is derived from seawater, of which we have plenty; and thirdly there are few radioactive components to be disposed of. It sounds too good to be true – so why haven't we done it yet?

It was a question I put one evening to a gathering of Europe's fusion researchers at a dinner in a hotel near Oxford, UK. Many answers came forth, money and political commitment chief among them, and then one delegate said that commercial fusion reactors could be as far away as the first pop hits of the Beatles. Is that good or bad, I wonder? Prophetically, one of the great Russians pioneers of fusion physics said: 'We will not harness the potential of fusion until it becomes a necessity.' Europe will still have sufficient fossil-fuel reserves to maintain its energy requirement at its present level well into the twenty-first century. But what then?

Fusion power offers the potential of an almost limitless source of energy for future generations, but it also presents some formidable scientific and engineering challenges. At the start of the twenty-first century, two billion of those six billion people live without electricity. Energy poverty equates with financial poverty and developing energy supplies to the world's poorest is vital if their lives are to improve. Every generation has its taboo and ours is this: that the resource on which our lives have been built is running out. We don't talk about it because we cannot imagine it. We are a civilisation in denial.

At a fusion conference in the United States a few years ago, some of the delegates got together for a meal. They were a true international cross-section. All five were in their fifties. When they sat down to dinner one of them raised a toast. 'May your children live to see fusion power,' he said. Another replied, 'May our grandchildren live to see fusion power.' A third then raised his glass, 'May our children have electricity.'

German Gustav Sporer was a school teacher who only started observing the Sun at age 36.

CHAPTER TWENTY

'Our summers are no summers'

In 1643, as the English Civil War raged, a preacher told the House of Commons in London that 'these are days of shaking ... and this shaking is universal: the Palatinate, Bohemia, Germany, Catalonia, Portugal, Ireland, England.' Later the Tsar of Muscovy was told, 'The whole world is shaking, and the people are troubled.' There were upheavals in the Ottoman Empire, Portugal, Sicily, Spain, Sweden and the Ukraine as well as Brazil, India and China in the world beyond. Ralph Josselin, an English vicar, wrote, 'I find nothing but confusions.' Josselin looked for God's inscrutable purpose, but others looked elsewhere. The Italian historian Majolino Bisaccioni suggested that the wave of revolutions might be due to the influence of the stars, but Jesuit and astronomer Giovanni Battista Riccioli speculated that fluctuations in the number of sunspots might be to blame.

As we have seen, something was happening to the Sun at that time and astronomers and philosophers of the time knew it. The early telescopic observers revealed numerous sunspots, but from the 1640s to the early eighteenth century they noted with alarm their almost total absence. When Giovanni Cassini, the director of the Paris Observatory, saw one in 1671, the *Philosophical Transactions of the Royal Society of London* at once reported it, adding a description of what sunspots were, because the last one had faded away a decade before and readers might have forgotten what they were. It appears that fewer sunspots were observed between 1645 and 1715 than appear in a single year nowadays. The English poet Andrew Marvell wrote a poem in 1667 called

'The Last Instructions to a Painter':

Man to the Sun apply'd
And Spots unknown to the bright star descry'd;
Showed they obscure him, while too near they please ...
Through Optick Trunk the Planet seem'd to hear,
And hurls them off, e're since, in his Career.

In May 1684 John Flamsteed, the first English Astronomer Royal, wrote, 'tis near seven and a half years since I saw one before they have been of late so scarce how ever frequent in the days of Galileo and Scheiner'.

The same effect occurred with the aurora borealis, the northern lights. After 1640 they became so rare that when Edmund Halley, England's second Astronomer Royal, saw them in 1716 he was overjoyed and wrote a paper about it. It was the first time he had seen them in almost 50 years of observation. Likewise, the brilliant corona – the sun's atmosphere – which is nowadays visible around the Moon during a total solar eclipse also disappeared: descriptions by Asian and European astronomers between the 1640s and 1700s mention only a pale ring of dull light, reddish and narrow. The energy of the Sun appeared to have diminished and the earth felt a chill.

Little Ice Age and Mediaeval Warm Period

There are many ways to estimate the Earth's climate over the past 1000 years or so. Scientists have been ingenious in finding what they term 'proxies', measurements that in some way reflect the climate. These proxies are an impressive spread of phenomena: tree rings, stalagmites, boreholes, pollen counts, sea sediments, coral, ice cores and mountain glacier deposits. The remarkable thing about all of them is that they agree that in the past millennium or so there have been two global climate anomalies: the so-called Little Ice Age, which stretches from 1300 to 1900 AD, and the Mediaeval Warm Period, from 800 to 1200 AD. Especially notable was the mild weather that had encouraged mediaeval Vikings to establish colonies in Greenland, colonies that endured for centuries only to perish from starvation in the Little Ice Age. Such global climate swings had been noted in the 1960s. British climatologist Hubert Lamb wrote:

> Multifarious evidence of a meteorological nature from historical records, as well as archaeological, botanical and glaciological evidence in various parts of the world from the Arctic to New Zealand ... has been found to suggest a warmer epoch lasting several centuries between about AD 900 or 1000 and about 1200 or 1300 ... Both the 'Little Optimum' in the Middle Ages and the cold epochs (i.e., 'Little Ice Age'), now known to have reached its culminating stages between 1550 and 1700, can today be substantiated by enough data to repay meteorological investigation.
>
> The commonest indications from very diverse types of evidence are that the prevailing temperatures in many parts of the world at least between 1000 and 1200 ... were 1–2 degrees C above the present values ... the medieval warm epoch and the subsequent cold centuries, the so-called 'Little Ice Age,' are confirmed.

Many French villages record the date at which the local grapes became ready to harvest. They show that, starting in the mid-seventeenth century, they ripened one or two weeks later than before. Cereal production also slumped in the mid-seventeenth century.

Everyone felt the cold and knew in the back of their mind that things were worse than they were a generation or two previously. Tidal rivers like the Thames froze over for long periods in the winter, allowing the famous Frost Fairs. Travellers in Scotland said that the main peaks of the Grampians and Cairngorms retained their snow all year. Surges of cold water southward from the polar regions ruined the cod industry off Iceland. As the fish's kidney reacts badly to water colder than 2 degrees Celsius, there was no cod between 1675 and 1750. In 1695 it was reported that an Inuit was found in his kayak in the river Don in Aberdeen. The same year the canals of Venice froze in the winter; Galileo would have been surprised. The English preacher John King wrote, 'Our years are turned upside down, our summers are no summers; our harvests, no harvests.'

The Little Ice Age affected Europe at just the wrong time. In response to the more benign climate of the Medieval Warm Period, Europe's population may have doubled. More people married and most did so earlier, giving birth to six or seven children despite, or perhaps because of, infant mortality being high. But in the mid-seventeenth century demographic growth stopped and in some areas fell, in part due to the reduced crop yields. Bread prices doubled and then quintupled. Buying bread now absorbed almost all a family's income, in turn causing the demand for manufactured goods to collapse, and resulted in unemployment.

High prices and reduced incomes forced many couples in Europe to marry later and the average age of brides rose from teenagers in the later sixteenth century to 27 or 28 in the mid-seventeenth century, reducing the birth rate. Hunger weakened the population. The English philosopher Thomas Hobbes argued in 1651 that 'the life of man [is] solitary, poor, nasty, brutish, and short'.

But what had happened to the Sun during this period and was it to blame, partially or wholly, for the Little Ice Age and the Mediaeval Warm Period?

Scientists, and journalists as well, should always be worried about assumptions. All too often in the history of science an unquestioned assumption, something taken as given, has turned out to be wrong, rendering all the implications that flow from it irrelevant or at least reduced in importance. I have those thoughts in the back of my mind when I watch or read the ever-growing number of reports about global warming. Mankind has undoubtedly been churning an unprecedented amount of greenhouse gases into the atmosphere through the burning of fossil fuels, and those gases will trap the heat from the Sun and warm our planet.

Nevertheless, two things nag at the back of my mind. One is that, working in a daily news environment myself, I know the way some (though not all) journalists produce such reports. They frequently tend to be, let me say, 'fashionable', with shallow scholarship and reflecting the public mood of the time rather than the full spectrum of the science concerned. Anything that happens to our climate, and many other things besides, is too often automatically put down to global warming. Frequently local news reports in southern England talk of flooding in the region, which is, they say, due to global warming. Weather is unpredictable and variable, extremes occur all the time, but these days many are all too keen to interpret those extremes as obviously global warming. That is my first worry. My second is that seldom do any of these reports, be they from news agencies, politicians or environmental pressure groups, ever mention the Sun.

We have seen that there are global climate swings that have occurred in the past that are the equal of the global warming claimed today. In those times we did not burn much fossil fuel, so mankind had nothing to do with those past examples of global warming or global cooling. The big question we should be asking is whether the Sun's activity could explain the global warming seen in the twentieth century. By the 1990s some believed that we had a tentative answer: solar variation could well have been responsible for some past fluctuations, but future manmade greenhouse effect warming would outweigh the solar effects. Or would it?

Solar–climate connections

When scientists first began to consider climate change, their thoughts did initially turn to the Sun. By the end of the nineteenth century a small group of them were pursuing the question of how solar variability might relate to short-term weather cycles, as well as long-term climate changes. But attempts to correlate weather patterns with the sunspot cycle were not successful because of inaccurate and disparate weather data, and a lack of good statistical analysis techniques. At the end of the nineteenth century, most meteorologists firmly believed that the Earth's climate was stable overall. Some scientists persisted in coming up with one or another plausible cycle of dry summers or cold winters or whatever, in one or another region, repeating periodically over intervals ranging from 11 years to several centuries, but they remained oddities, pushed to the margins and not taken as indicative of anything larger at all. Confusion persisted.

I do not know if Ellsworth Huntington, professor of economics at Yale University in the twentieth century known for his studies into economic growth and geography, ever read Agatha Christie. If he did he may have come across the passage where Miss Marple says: 'Any coincidence ... is always worth noticing. You can throw it away later if it is only a coincidence.' Drawing on work by others, Huntington concluded that high sunspot numbers meant storminess and rain in some parts of the world, resulting in a cooler planet. He said that the 'present variations of climate are connected with solar changes much more closely than has hitherto been supposed'. He went on to say that solar disturbances might explain the ice ages.

In the 1930s another advocate of a solar climate connection was Charles Greeley Abbot of the Smithsonian Astrophysical Observatory. His predecessor, Samuel Pierpont Langley, had established a programme of measuring the 'solar constant' that Abbot pursued for decades. By the early 1920s, he believed he had evidence that the Sun varied its brightness and that the constant was misnamed because his analysis showed large variations over periods of days, which he connected with sunspots passing across the face of the Sun. He found that over a period of years the Sun seemed brighter by nearly 1% and believed that this must influence the Earth's climate. Indeed, as early as 1913, Abbot announced that he could see a plain correlation between the sunspot cycle and cycles of temperature on Earth. So a rather combative Abbot told the public and his fellow scientists that solar studies would bring improvements in weather prediction. He and a few others at the Smithsonian pursued the topic into the 1960s, convinced that sunspot variations were a main cause of climate change, but other scientists were sceptical as Abbot's solar constant variations were at the edge of detectability if not beyond. You needed the eye of faith to be convinced by them.

By the 1920s and 1930s there was a lot of weather data to play with and inevitably, people found correlations between sunspot cycles and selected weather patterns that they thought could be used to forecast the weather. An example was a forecast that there would be a dry spell in Africa during the sunspot minimum of the early 1930s. When that proved to be wrong, a meteorologist said, 'The subject of sunspots and weather relationships fell into disrepute, especially among British meteorologists who witnessed the discomfiture of some of their most respected superiors.' Even in the 1960s it was said, 'for a young researcher to entertain any statement of sun–weather relationships was to brand one a crank.'

Specialists in solar physics felt much the same. One said, 'Purported connections with ... weather and climate were uniformly wacky and to be distrusted ... there is a hypnotism about cycles that ... draws all kinds of creatures out of the woodwork.' So it was that generally, by the 1940s, most meteorologists and astronomers had abandoned the quest for solar cycles in the weather. But was there really something profound lurking somewhere in the data?

During the 1920s, a few scientists developed simple models of the Earth's climate system that suggested that even a modest change in solar radiation might set off an ice age, by causing changes in the polar ice. A leading British meteorologist, Sir George Simpson, believed that the sequence of ice ages showed that the Sun is a variable star, changing in brightness over a cycle some 100,000 years long. He told the Royal Meteorological Society in a Presidential address in 1939:

> There has always been reluctance among scientists to call upon changes in solar radiation ... to account for climatic changes. The Sun is so mighty and the radiation emitted so immense that relatively short period changes ... have been almost unthinkable.

But he added that none of the other causes proposed to explain the ice ages was at all convincing, and that this 'forced a reconsideration of extraterrestrial causes'.

Much later the eminent astrophysicist Ernst Öpik wrote that none of the many explanations proposed for ice ages was convincing, so 'we always come back to the simplest and most plausible hypothesis: that our solar furnace varies in its output of heat'. So it was that a kind of paradigm was established. In the 1950s textbooks listed various explanations of ice ages and other long-term climate changes, ranging from volcanic dust to shifts of ocean currents, but in the end they usually called on long-term solar variation as the likely cause. As one US Weather Bureau expert put it, 'the problem of predicting the future climate of Planet Earth would seem to depend on predicting the future energy output of the sun.'

Nevertheless, some continued to pursue the hints that minor variations in the sunspot cycle influenced the weather. In 1949 meteorologist H. C. Willett investigated long-term variations of large-scale wind patterns and changes in the numbers of sunspots and declared that they were 'the only possible single factor of climatic control which might be made to account for all of these variations'. Some thought they detected a sunspot cycle signal in the advance and retreat of mountain glaciers. Willett admitted that 'the physical basis of any such relationship must be utterly complex, and is as yet not at all understood'.

All this would have pleased Miss Marple. Were they 'just' coincidences? Willet said that perhaps climate changes could be due to 'solar variation in the ultraviolet of the sort which appears to accompany sunspot activity'. It had been pointed out before that ultraviolet radiation from the explosive flares that accompany sunspots would heat the ozone layer high in the Earth's atmosphere, and that might somehow influence the circulation of the lower atmosphere. Perhaps the scientists were onto something?

In the 1950s and 1960s, rockets climbed above the Earth's atmosphere for the first time and managed to measure the Sun's ultraviolet radiation. They found that it did intensify during high sunspot years, but because it does not penetrate below the stratosphere – the upper regions of Earth's atmosphere – meteorologists were sceptical that changes in the thin stratosphere could affect the layers below.

Some scientists suggested that maybe weather patterns were affected by the electrically charged particles that the Sun sprayed out as the 'solar wind'. They also proposed that at times of high sunspot activity the solar wind pushes out a magnetic field that tends to shield the Earth from the cosmic rays that rain down from deep space. When these rays penetrate the upper reaches of the atmosphere, they expend their energy producing sprays of charged particles – so more sunspots would mean fewer of these particles. Either way, there might be an influence on the weather.

In 1961 Minze Stuiver, a geoscientist at the University of Washington, was concerned about variations in the amount of radioactive carbon-14 found in ancient tree rings. Carbon-14 is generated in the atmosphere by cosmic rays from the far reaches of the universe. Stuiver noted how changes in the magnetic field of the Sun would change the flux of cosmic rays reaching the Earth. He had followed this up in collaboration with the carbon-14 expert Hans Suess, confirming that the concentration of the isotope had varied over past millennia. In the 1960s a few scientists tried correlating the new data with weather records, in the hope that carbon-14 variations may supply conclusive evidence regarding the causes of the great ice ages. They focused on the Little Ice Age, which had been a time of relatively high carbon-14, pointing to low solar activity. Suess noticed that the same centuries showed a low count of sunspots. In short, fewer sunspots apparently made for colder winters. A few others found the connection plausible, but to most scientists the speculation sounded like just one more of the countless, unsupportable correlations that had been rife over the previous century.

The 1970s also brought controversial claims that weather data and tree rings from various parts of the American West revealed a 22-year cycle of droughts, presumably driven by the solar magnetic cycle. Scientists were beginning to understand, however, that the planet's climate system could go through purely self-sustaining oscillations, driven by feedbacks between ocean temperatures and wind patterns. The patterns cycled quasi-regularly by themselves on timescales ranging from a few years (like the important El Niño Southern Oscillation in the Pacific Ocean) to several decades. That might help to explain at least some of the quasi-regular cycles that had been tentatively associated with sunspots.

'We had adopted a kind of solar uniformitarianism,' solar physicist Jack Eddy explained in retrospect. 'As people and as scientists we have always wanted the Sun to be better than other stars and better than it really is.'

In 1975, Robert Dickinson, of the National Center for Atmospheric Research in Boulder, Colorado, reviewed the American Meteorological Society's official statement about solar influences on weather. He said something that to me was quite remarkable for a scientist, that such influences were unlikely, for there was no reasonable mechanism in sight – except maybe one. Perhaps the electrical charges that cosmic rays brought into the atmosphere somehow affected how aerosol particles coalesced. Perhaps that somehow affected cloudiness, since cloud droplets condensed on the nuclei formed by aerosol particles. This was just piling speculation on speculation, Dickinson pointed out. Scientists knew very little about such processes, and would need to do much more research 'to be able to verify or (as seems more likely) to disprove these ideas'.

Dickinson had obviously not read Agatha Christie. It does not matter that a scientist may not know what the explanation for a coincidence is, a coincidence may be indicative of something deeper that may not have a current explanation – this is science, after all. The Sun's activity record correlated with global climate changes on earth and it could not be dismissed as unrelated because the underlying mechanism that connected them was not known. Shades of Lord Kelvin again. But for all his scepticism, Dickinson had left the door open a crack. It was now at least conceivable that changes in sunspots could have something to do with changes in climate.

So it was in 1976 that Jack Eddy tied all the threads together in a paper that became famous. Eddy found that the Sun was by no means as constant as astrophysicists supposed.

'As a solar astronomer I felt certain that it could never have happened,' Eddy later recalled. But hard historical work gradually persuaded him that the early modern solar observers like Galileo and Scheiner were reliable – the absence of sunspot evidence really was evidence of an absence.

Eddy commented:

> [It was] one more defeat in our long and losing battle to keep the Sun perfect, or, if not perfect, constant, and if inconstant, regular. Why we think the Sun should be any of these when other stars are not is more a question for social than for physical science.

But his announcement of a solar–climate connection nevertheless met with the customary scepticism. He pushed his arguments, stressing the Little Ice Age, which as we have noted he called the 'Maunder Minimum' of sunspots. The name he chose emphasised that he was not alone with his evidence. Eddy warned that in our own times, 'when we have observed the Sun most intensively, its behaviour may have been unusually regular and benign'.

The sound of a Stradivarius

For some the sound of a Stradivarius violin is like no other. It is lively, it flickers and trembles like candlelight, an indefinable combination of brilliance and darkness, a pure tone. Probably only a very few experts could immediately tell a Stradivarius and in blind tests modern instruments have been said to sound better (to be fair, there are also differences in construction and sometimes older instruments are not at their best when played in the modern style, as they were not meant to be). These factors, however, do not stop the greatest violinists regarding them as the gold standard. There is a special thrill in hearing a violin concerto when it is played on a Stradivarius.

The world's most famous violin is the so-called Messiah, made in 1716 in Cremona by Antonio Stradivari. It lives in the Ashmolean Museum in Oxford and access to it is strictly limited. It is estimated to be worth some £15 million. It is said that Stradivari so admired its delicate lines that he jealously guarded it right up to his death in 1737, without it ever having been played.

During his lifetime, it is estimated that Stradivari made 1100 instruments – violins, guitars, violas and cellos – of which about 600 survive. But what is it that makes these instruments so prized?

The popular belief is that the Cremonese artisans of the late seventeenth to eighteenth centuries had a 'secret ingredient' (or undocumented technique) that gave them their famed sound. Candidates include the use of a special varnish, chemical treatment of the wood, 'cooking' or drying the wood and the use of very old wood from historical structures. But the answer may be found in the Sun and sunspots.

Stradivari was born at the start of the Maunder Minimum and the Little Ice Age. It has been speculated that the long winters and cool summers of the period produced wood that provides the unique, rich sound of the Cremonese instruments. Violin makers have always known that the selection of the wood makes all the difference to the sound of the instrument. Maple is preferred for the back, ribs and neck, while spruce for the top.

Stradivari and other eminent Italian violin makers of the time who worked in Cremona probably used the nearby forests of the southern Italian Alps as their source of their spruce. Perhaps the trees growing during Stradivari's lifetime experienced a unique set of environmental conditions that has not occurred since.

Recent evidence

The Sun's influence on climate is a puzzle indeed, as our tour of the history of research into the connections between solar activity and climate has shown. But there are recent developments, some thought-provoking, some significant and to some people highly unpopular.

The United Nations' Intergovernmental Panel on Climate Change (IPCC) claims that human activities are responsible for nearly all the Earth's recorded warming during the past two centuries. At issue is what is commonly referred to as the 'hockey stick', a widely circulated image that depicts a 1000-year period where temperatures remained relatively constant, followed by the last 100-plus years where temperatures have shot upwards. It was created by researchers Michael Mann of the University of Virginia and Phil Jones of the University of East Anglia, and it is used by the IPCC and environmental activists as proof of human-induced global warming.

It is in light of this that people like Sir David King, scientific adviser to Her Majesty's Government in the UK, make statements such as 'by the end of this century, the only continent we will be able to live on is Antarctica'. In 2000 the IPCC said of the 'hockey stick', 'Reconstructions of climate data for the past 1,000 years ... indicate that the warming over the past 100 years was unusual and is unlikely to be entirely natural in origin.' But there is a problem with the model.

It now seems that the researchers may have made errors in the collection and use of data from multiple sources. In some cases they have used obsolete data, made incorrect calculations, associated data sets with incorrect geographical locations, used incorrect statistics and in some cases underestimated the errors. If that were not bad enough, they removed long time period trends, such as the Mediaeval Warm Period and, crucially, the Little Ice Age.

There is also the problem of tree growth, on which the temperature curve rests heavily. Many researchers have noticed that trees at high latitudes have been showing unusual behaviour in recent decades, displaying a declining density of growth rings that is not related to temperature. Experts in tree growth do not have an explanation.

One of the researchers of the IPCC curve published a retraction in the June 2004 issue of *Geophysical Research*, in which he admits underestimating the temperature variations indicated by the proxy data since 1400 by more than one third, which accounts for why he missed the Little Ice Age. Strangely, the researcher still argues that this considerable error has no impact on his conclusions.

Furthermore, other researchers have been unable to reproduce the rising part of the hockey stick using common statistical techniques, or even employing the same techniques as Mann and Jones. So are the claims that humans have caused tremendous warming over the last 100 years and that the 1990s was the warmest decade untenable? Five independent research groups have uncovered problems with this reconstruction, calling into question all three components of the hockey stick. Perhaps it is now broken.

In 2001 Sallie Balliunas and Willie Soon of the Harvard Smithsonian Center for Astrophysics took a look at the IPCC data. Baliuanas has a distinguished track record looking for the equivalent of our Sun's sunspot cycle occurring on other stars. They concluded that the global warming observed over the past century is not unusual and 'is the result of natural factors in climate change'. They added that concerning human activities causing the recent global warming, the 'scientific evidence clearly indicates that this is not the case'.

I recall going to many science conferences in the late 1980s and 1990s. At some of them certain prominent scientists repeatedly said that the first signs of manmade global warming had not yet emerged but they must, just look at the quantity of greenhouse gases we are churning out into the atmosphere. Then, when the first global temperature measurement poked its head above the usual statistical fluctuation, they leapt for joy – there it is, manmade global warming, they said. Something nags at me about this. I think they may have prejudged the data, saw what they expected and interpreted it as they wished. Perhaps they were prejudiced.

Whether global warming is a reality and whatever the human contribution to it is, our predictions of the future will be irrelevant if we do not take heed of what the Sun is doing. The Sun may make things warmer or colder in the forthcoming century and we don't really know which. But there are clues.

Just over a decade ago a provocative finding was reported. Northern-hemisphere temperatures over the past 130 years correlated surprisingly well with the length of the sunspot cycle (which varied between 10 and 12 years). The report said that the late twentieth-century rise in global temperature might be entirely due to increased solar activity. Critics said that the new finding sounded like the old discredited sunspot work and that if you look hard enough you were bound to turn up something. To be fair, later work did not support the correlation between solar cycle length and increases in global temperature.

In 1997 a pair of Danish scientists drew attention to a possible mechanism for the link. Scanning a huge bank of observations compiled by an international satellite project, they found that global cloudiness increased slightly at times when the influx of cosmic rays was greater. Weaker solar activity resulted in more clouds, just like the tentative suggestions of the 1970s. When the cosmic rays hit the Earth's atmosphere they not only produced carbon-14, but also sprays of electrically charged molecules. Perhaps they were promoting the condensation of water droplets on aerosol particles, thus producing extra cloudiness.

New data on the cosmic ray flux as recorded in the beryllium-10 content of deep ocean sediments suggests that cosmic rays have something to do with ice ages. It implies a link between the number of cosmic rays arriving on Earth and the glacial cycles. Beryllium-10 is produced when cosmic rays interact with particles in the Earth's atmosphere and then falls to the ground, where it is stored in ice or ocean sediments.

The standard so-called insolation model of glacial cycles was first put forward by the Serbian astrophysicist Milutin Milankovitch in 1912. He proposed that ice ages were caused by small variations in the amount of sunlight hitting the Earth due to a very gradual cyclic change in the shape of the Earth's orbit around the Sun. However, while the insolation model can explain a glacial cycle with a period of 41,000 years that is seen in paleoclimatic data, it predicts a 400,000-year cycle that is not. Moreover, it cannot explain a 100,000-year cycle that is also present.

Other recent studies are reviving the old idea that increased ultraviolet radiation during times of higher solar activity might affect climate by altering stratospheric ozone. While the total radiation from the Sun in optical light was nearly constant, instruments on rockets and satellites have found that the strength of the Sun's ultraviolet light can vary by several percent over a sunspot cycle and today's computer models suggest that even tiny variations could make a difference.

Whatever the exact form solar influences take, I sense a growing willingness to give the Sun more credit for climate change than it has received in the politically correct 'it's all mankind's fault' times.

A 1999 study reported evidence that the Sun's magnetic field had strengthened greatly since the 1880s. Could it be that this has something to do with the temperature rise over the same period? In particular, the warming from the 1880s to the 1940s had come when solar activity had definitely been rising, while the build-up of greenhouse gases had not.

There is evidence that the flickering Sun may have caused rapid climate change in the more distant past. A 200-year cold snap some 10,300 years ago seems to have coincided with a decline in the Sun's activity, according to Swedish researchers. They came to this conclusion after looking at sediments in Lake Starvatn on the Faroe Islands and in the Norwegian Sea, growth rings in ancient German pine trees, and chemical analysis of ancient ice from the Greenland ice sheet.

A pertinent set of data has been collected by scientists at Armagh Observatory in Northern Ireland. They produced a unique 200-year long series of weather measurements that may show that the Sun is chiefly responsible for global warming. These weather observations, made almost every day for over two centuries, are the longest climate archive available for a single site in Ireland. The observations began in 1795, a few years after the Armagh Observatory was founded. Temperature, pressure and later rainfall have been measured every day, with the exception of a period around 1825.

When analysed, the data allow the average temperature at Armagh to be calculated to an accuracy of 0.1 degree Celsius per decade. What makes the data so useful is that the site of the observatory has not changed all that much in 200 years. Other weather stations have been engulfed by towns and cities that make the long-term reliability of their data questionable. In all that time the Armagh meteorological instruments have been moved only about 20 metres.

The researchers point out that the mean average temperature at Armagh seems to be related to the length of the Sun's activity cycle. They found that it gets cooler when the Sun's cycle is longer and that Armagh is warmer when the cycle is shorter.

And consider this. A significant recent analysis suggests that the Sun is more active now than at any time during the past 1100 years. It comes from a study by Swiss and German researchers, who say that sunspots are at a millennial high. Dr Sami Solanki, the director of the renowned Max Planck Institute for Solar System Research in Göttingen, says that the Sun is in a changed state. It is brighter than it was a few hundred years ago and this brightening started relatively recently – in the last 100 to 150 years. He believes that the brighter Sun and higher levels of greenhouse gases, such as carbon dioxide, have both contributed to the change in the Earth's temperature, but that it is impossible to say which had the greater impact.

Climate change is a reality. Just over 10,000 years ago Oslo was under a kilometre of ice and 7000 years ago the Sahara was green and lush. Whatever the impact mankind is making on its environment, the Sun is influencing it as well and is capable of the current warming all on its own. Go listen to the sound of a Stradivarius.

Sir Arthur Eddington wanted to understand 'so simple a thing as a star'.

CHAPTER TWENTY-ONE

Shelter from the storm

The core of the Sun is at a temperature of about 15 million degrees. It is its powerhouse, where nuclear fusion takes place, generating energy when protons fuse. We cannot see this process happening as it takes place at the centre of a ball of gas some 1.4 million km across. We can only see what is happening at the surface.

Well, not exactly. There is a way to peer deep inside the Sun, right to its very heart, and when we first did this it threw up a puzzle that for a while threatened to topple all the understanding we had spent centuries building up, about our star and indeed about nuclear physics itself. It all happened because of a ghostly particle called the neutrino.

The neutrino is a remarkable subatomic particle. It hardly interacts with matter at all, and if it were possible to construct several million million kilometres of solid lead it would be able to pass through that without being stopped. It has been called the nearest anything can get to nothing while still being something. And yet for years this tiny, elusive particle lay at the heart of a great mystery concerning our Sun. A mystery that was solved using the world's biggest tank of water.

The existence of neutrinos was first proposed by the Italian-American physicist Wolfgang Pauli in a 1930 letter to his physics colleagues as a 'desperate way out' of a physics problem. Beta decay is a process that an unstable atom can use to become more stable. Pauli thought that he understood one form of beta decay when a neutron in the nucleus of an atom turns into a proton and an electron (remember that neutrons and protons comprise an atomic nucleus that for a normal atom is surrounded by an electron 'cloud'). The proton stays in the nucleus while the electron is ejected. But the energies did not match up. When Pauli observed such examples of beta decay, he noticed that the electron was ejected without enough energy, so he hypothesised that there was another particle he could not see that was created when the neutron decayed that accounted for the missing energy. According to Pauli's hypothesis, which he put forward very hesitantly, neutrinos are elusive particles that escape with the missing energy in beta decay.

The mathematical theory of beta decay was developed by Enrico Fermi in 1934, in a paper that was rejected by the journal *Nature* because 'it contained speculations too remote from reality to be of interest to the reader'. Neutrinos from a nuclear reactor were first detected by Clyde Cowan and Fred Reines in 1956.

We have already seen that based on the work of Hans Bethe and his colleagues, we believe that four hydrogen nuclei are burned into a single helium nucleus, plus two positive electrons and two neutrinos, plus energy. This means that from the very core of the Sun, from its energy-generation regions, far beyond all sight, from which radiant energy takes a hundred thousand years to emerge, there are particles that escape as if the surrounding Sun were not there at all. If we could detect these solar neutrinos we would be able to look into the heart of the Sun and measure what the conditions were like there at the moment. We can calculate how many neutrinos are arriving at Earth direct from the Sun's core. The answer is about a hundred million million of them per square metre per second. Billions are passing through your body every second doing you no harm whatever.

In 1964 an experiment was proposed to try to detect neutrinos. It was suggested that a huge tank containing 100,000 gallons of cleaning fluid (perchloroethylene, which is mostly composed of chlorine) could detect the neutrinos from the Sun. When neutrinos collide with a chlorine nucleus they produce a radioactive isotope of argon. This happens very rarely, but if there is a large flux of them through the tank a few of them might collide occasionally. All you had to do was examine the tank every once in a while and measure the argon concentration, and you could estimate the neutrino flux passing through the tank from the Sun. Just a few atoms would be produced per week in a huge Olympic-sized swimming pool of cleaning fluid. The tank was set up 5000 feet underground in the Homestake Gold Mine in South Dakota. Underneath all that rock, the tank would be shielded from cosmic rays that could cause confusion. The solid rock was, of course, irrelevant to the neutrinos.

In 1968 it was announced that fewer neutrinos than predicted had been detected coming from the Sun. At first scientists thought that there was something wrong with the tricky experiment, but the discrepancy persisted. There really were far fewer neutrinos coming out of the Sun than theory predicted. So was the theory of energy production inside the Sun wrong? Or did something happen to the neutrinos after they were created in the Sun?

It seemed inconceivable that the theory was wrong. Perhaps the Sun's central temperature was slightly lower than we thought, which would affect the emerging neutrino flux. But even that would cause problems for our understanding of the stability and lifetimes of stars. Perhaps the most imaginative proposal was made by Stephen Hawking, who suggested that the central region of the Sun might contain a small black hole and that this could be the reason that the number of neutrinos observed was less than the predicted number.

But as early as 1969, two scientists working in Russia proposed that the discrepancy between standard theory and the first solar neutrino experiment could be due to the behaviour of the neutrinos themselves. Gribov and Pontecorvo suggested that neutrinos suffer from a multiple personality disorder in that they oscillate back and forth between different states or types. According to these scientists, neutrinos of different types are produced in the Sun. As they travel to the Earth from the Sun, the neutrinos oscillate between the easier-to-detect neutrino state and the more difficult-to-detect neutrino state. The chlorine experiment only detects neutrinos in the easier-to-observe state, so the measured flux was low.

Over the next few decades physicists built a series of underground detectors designed to detect neutrinos. This time they did not use cleaning fluid but water. They lined their tank with light detectors that could pick out any flashes of light from the water. Occasionally a neutrino would cause a flash in the tank, which would be detected. In 1986, Japanese physicists used such a huge tank of water designed to measure the stability of matter. The experiment, called Kamiokande, in the Japanese Alps detected solar neutrinos and confirmed that the neutrino rate was less than predicted by standard physics and standard models of the Sun. So far neutrinos produced in the centre of the Sun have been detected in five experiments.

But neutrinos are also produced by the collisions of cosmic ray particles with other particles in the Earth's atmosphere. In 1998, the Super-Kamiokande team of experimentalists announced that they had observed oscillations among atmospheric neutrinos.

So it now seems that the mystery that once threatened to undermine our understanding of the Sun has gone away. As for the scientist who had the audacity to fill an underground tank with chlorine cleaning fluid, Raymond Davis Jr. won the 2002 Nobel Prize for Physics.

Solar radiation

Neutrinos pose no problem for astronauts, but the Sun produces many other forms of radiation that do.

Imagine ... It is 1972 and the last planned manned mission to the Moon, *Apollo 17*, is cancelled after the death of the crew of *Apollo 16* when returning to the Earth after their Moon landing. The United States and the world are stunned by the loss of three brave astronauts who succumbed to radiation in space and became too ill to control their spacecraft during its return.

SHELTER FROM THE STORM

It could have happened. From time to time the Sun unleashes monster flares that could put astronauts travelling to the Moon and beyond in great jeopardy. The plan for the Apollo lunar missions, which placed 12 people on the lunar surface between 1969 and 1972, called for low-altitude Earth orbits and then a rapid transit to the Moon. While in low Earth orbit the Earth's magnetic field shelters astronauts from harmful solar radiation. The worry on all manned lunar missions was a sudden explosion on the Sun and the release of vast amounts of radiation while the astronauts were beyond the Earth's protective custody.

It was to keep a watch for such unexpected events that before and during the Apollo missions the Sun was maintained under close scrutiny. The Solar Particle Alert Network monitored solar flares. A prognosis of the expected radiation dose the astronauts might experience was prepared and continually updated by radiation environment specialists sitting at a console in the Apollo Mission Control Center in Houston. The dose estimates were given to the medical officer, who advised the flight director of the radiation effects to be expected. Onboard the Apollo spacecraft there were radiation monitors and each astronaut carried a personal dosimeter, which measured his accumulated skin dose. A portable dose rate meter was also carried in the lunar module and onto the lunar surface.

In August 1972 preparations were well underway for the *Apollo 17* mission to the Taurus–Littrow region of the Moon in December. It was to be the final Moon landing, as *Apollos 18–20* had been cancelled. The previous Apollo mission, in April to the plains of Descartes, had been a resounding success. But that summer month something happened that literally frightened mission controllers and made them realise, if they needed any reminding, just how dangerous was a flight to the Moon.

The August 1972 solar flare was a monster that above the Earth's protective magnetic sheath delivered unshielded radiation doses of 20,000 REM. It is thought that flares with such particle fluxes occur only about one day per decade, but the implications of any crew being caught out in such an event are terrifying. The

REM is a standard measure of radiation exposure. On a 'normal' Apollo mission to the Moon the worst the astronauts would expect is about 1 REM, which comes from transit through the Van Allen radiation belts that surround the Earth.

The US Space Agency specifies a 30-day exposure limit of 25 REM and the terrestrial occupational limit for radiation workers is 5 REM in a year. The average lethal dose is 450 REM, which means that any astronaut in space or on the Moon's surface at the time of the August 1972 solar flare would have suffered a lethal exposure if they could not find protection. But what would have happened if two astronauts had been working on the lunar surface and a third had been in orbit around the Moon that August day and had suffered an exposure of almost 20,000 REM?

Between 25 and 100 REM there is an increased probability of leukaemia during the unfortunate person's lifetime and damage to their germ cells could mean they would pass on genetic defects to their offspring. Between 100 and 250 REM nausea and vomiting occur within hours and there is a high incidence of leukaemia in a shortened life. Untreated, it can lead to death. Between 250 and 750 REM there is nausea and vomiting after about an hour along with acute shock. Death is likely within a month if untreated. There will be cancers, cataracts, significant life shortening and sterility. Between 750 and 2000 REM there is nausea and vomiting within an hour; unconsciousness, then a temporary return to consciousness, but death within a week. Above 2000 REM the crew would be unconscious in minutes, with a brief recovery, then death. In the one recorded instance of 10,000 REM exposure death occurred within 38 hours, incapacitance much sooner.

The fact is that if we are to explore the Moon and go onto Mars, then some day astronauts will have to deal with and survive a big solar flare. Flares large enough to require shielding occur perhaps several times a year. There have been many studies in relation to the Moon and all follow the principle of placing as much mass as possible between the astronaut and the Sun. Perhaps we must position the first manned outposts on the Moon near the lunar poles, where there are regions that sunlight can never get to and where a shelter on the poleward side of a crater wall or in a permanently dark polar crater would provide protection. If we build bases elsewhere on the Moon we had best bury them in as much lunar dirt as we can manage.

Long-traverse pressurised lunar rovers present special problems, since the maximum solar flare warning is currently just three hours, so the time needed for astronauts to reach shelter would be likely to exceed the onset of the flare. Designs have called for such vehicles to carry an inflatable shelter that can be buried under lunar dirt.

Outside the Earth's magnetic shield (the magnetosphere), an interplanetary astronaut caught by such a flare would also need some sort of special protection. The Apollo astronauts reported seeing occasional flashes in their eyes during the Moon missions and close examination of their helmets revealed microscopic holes where high-velocity particles (hydrogen and helium nuclei) had tunnelled through the spacecraft, through their helmet and through their heads! Some of these particles were cosmic, coming from outside our solar system, but others are thought to have been solar in origin.

The traverse to Mars will probably take the best part of a year. Even if we used new-generation nuclear propulsion, yet to be designed and tested, the flight time to Mars would be reduced to perhaps four months. This means that there would have to be some form of storm shelter in the depths of the spacecraft, surrounded by as much hardware as possible. Perhaps the shelter would be at the end of the spacecraft with the vehicle turning to face the Sun, placing the maximum amount of material between the Sun and the astronauts.

One thing is for sure: wherever astronauts roam in our solar system they will sometimes have to hide from the Sun.

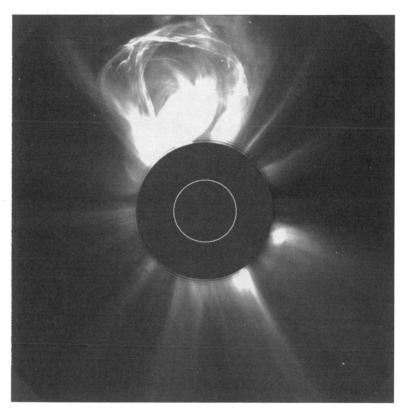

A storm from the Sun heads into interplanetary space.

CHAPTER TWENTY-TWO

The rescue of SolarMax

Since before the birth of the space age astronomers had always wanted to study the Sun from space, as it was clear that the Earth's atmosphere prevented much radiation from reaching observatories on the Earth's surface. Instruments could be carried up into the upper atmosphere by balloons and small rockets could carry a telescope into space for a few minutes, but what scientists really wanted was a satellite either in Earth orbit or in orbit around the Sun.

The observation of the Sun from space has revolutionised the study of our star. Probes have travelled close to it (but not too close) and high above it. They have sampled its solar wind and monitored blast waves as they travel through the solar system, they have peered deep into the heart of sunspots and watched as flares rip magnetic fields and spill superheated gas. Today, any day, you can use the Internet to obtain current images of the Sun in white light and in X-rays. You can look at ultraviolet images, and such things as magnetograms and dopplergrams as well as the X-ray flux. All theses are available direct from ground-based observatories and satellites. What is more, a fleet of satellites around the Earth keep a close eye on the Earth's 'space weather', so you can also look in real time at the images of the auroral oval above the north and south poles. Solar and solar-terrestrial data is now available to all. But it was not always the case.

In 1958, shortly after the launch of *Sputnik*, the United States had just placed its first satellite into orbit, *Explorer 1*, which discovered belts of radiation surrounding the Earth – the so-called Van Allen belts. While the Americans were drawing up plans to catch up with the Soviets, whose first satellites were much larger than anything the US could put into space, they began studying the idea of a probe to explore the planets and between the planets. The problems were enormous: no space vehicle had ever transmitted from distances of tens of millions of kilometres, and no satellite had ever endured months of solar irradiation. There were several possible options. Venus was reachable in about 150 days, departing in June 1959. A trip to Mars would take about 250 days, starting in October 1960.

Meanwhile, the Soviet's first Moon probe, *Mechta* (Dream) was launched on 2 January 1959. Although it was due to strike the Moon, it missed by about 6000 km and became the first artificial planet. However, transmissions ceased after 62 hours at a distance of 600,000 km.

The Americans' plans for a true interplanetary probe were running into difficulties as launch windows came and went and no probe was ready. Eventually, it was decided that the mission – *Pioneer 5* – would be launched on a pseudo-trajectory to Venus. When it reached the orbit of Venus the planet was elsewhere, allowing the probe to test long-range transmissions and transmit data on the interplanetary environment. *Pioneer 5* was based on the *Explorer 6* satellite launched in August 1959. On board were two radio transmitters that could be activated on command, able to transmit data at three different speeds so as to adapt to the increasing distance from Earth. There were five scientific instruments.

Pioneer 5 left pad 17A of Cape Canaveral on 11 March 1960, becoming an artificial planet in an orbit that would have taken it between 148.4 and 120.6 million km from the Sun in a 312-day period. It made many remarkable discoveries, including that the Earth's magnetosphere was much wider than had been thought. But at the end of March a flare erupted on the Sun that was observed both on board *Pioneer 5* and on the Earth some 20 minutes later. During one flare high-energy protons were recorded whizzing past the probe, and at the same time it was noticed that the flux of galactic cosmic rays markedly decreased. In time this was to become a highly significant observation. During the periods of quiet Sun activity, *Pioneer 5* discovered the existence of an interplanetary magnetic field with a very low intensity. It transmitted useful data until 30 April 1960, after which contact was sporadic due to battery failure. The measurement of the probe's velocity enabled scientists to produce the best estimate then available for the Earth–Sun distance. In addition, it established a communications link with Earth from a record distance of 36 million km on 26 June 1960, which was the last day of transmission.

Following *Pioneer 5*'s good start, *Pioneers 6–9* (1965–68) demonstrated the practicality of spinning a spacecraft in order to stabilise it to simplify its control. Measurements made by these spacecraft greatly increased our knowledge of the interplanetary environment and the effects of solar activity on Earth. New information was gathered about the solar wind, cosmic rays, the structure of the Sun's plasma and magnetic fields and the nature of solar flares. During the Apollo Moon landings, NASA used the fleet of Pioneers to provide hourly updates on the Sun's activity to mission control.

Originally designed to operate in space for at least six months, the Pioneers have proved to be remarkably long-lived. *Pioneer 9* failed in 1983. *Pioneer 8* was last tracked successfully on 22 August 1996. *Pioneer 7* was last tracked successfully in March 1995. But it is *Pioneer 6* that holds the title of the oldest operating spacecraft ever. To mark its 35 years in orbit as the oldest extant NASA spacecraft, one last contact was successfully completed on the 70 meter Deep Space Station antenna at Goldstone, California on 8 December 2000. If we look again in 35 years' time it could still be transmitting.

The US Orbiting Solar Observatories (OSOs) were the earliest set of satellites designed specifically to study the Sun. They had their origin in early sounding rocket flights that showed the importance of getting above the Earth's atmosphere. Orbiting Solar Observatory 1 was launched on 7 March 1962 to study the Sun in the ultraviolet, X-ray and gamma-ray regions of the spectrum. Sun sensors connected to feedback systems on its upper segment were designed to keep the pointed instruments looking at the Sun. A lower, spinning section carried other instruments and rotated once every two seconds, allowing them to scan the solar disk and atmosphere. Among many observations by the battery of instruments on OSO 1 was that the Sun's corona had openings, now called coronal holes, which were interpreted as huge fast-moving bubbles rising through the corona. The eighth and last Orbiting Solar Observatory was launched on 21 June 1975.

Skylab

Skylab, the first US space station, was launched into orbit on 14 May 1973 as part of the Apollo programme. It was basically a converted Saturn rocket stage that had been pressurised and filled with equipment and living quarters, using up Apollo hardware that was left over from the lunar landing missions. This 91-tonne structure was 36 metres (four stories) high, 6.7 metres in diameter and flew at an altitude of 435 km. Three different Apollo crews manned *Skylab* during its nine-month orbit.

However, there was a time when it seemed that *Skylab* would never be manned. Shortly after launch it seemed to be a useless orbiting hunk, crippled by a series of malfunctions. During ascent to orbit its micrometeorite thermal shield tore loose, taking with it one of its solar wings. To make matters worse, the second wing jammed when it tried to deploy. *Skylab* seemed doomed as it was adrift without power and thermal protection.

Its first crew was to go up the next day, but had to wait as an audacious rescue plan was concocted. A parasol was designed and taken up with the crew. They eventually docked and entered the space station, undertaking spacewalks to erect the parasol and free the stuck solar panel. *Skylab* was saved.

It was manned for 171 days by three crews during 1973 and 1974. It holds a special place in the exploration of the Sun because it carried a converted lunar module with four solar panels added to make the Apollo Telescope Mount (ATM), which contained a series of telescopes that the astronauts used to take more than 150,000 images of the Sun.

There are many solar astronomers, especially those who have been working in the field since the days of *Skylab*, who have a special fondness for the mission. It produced a remarkable series of images, some of which made it into a book produced by NASA and compiled by Jack Eddy. Even today, so many years after they were taken, I still thrill when I turn the pages of this book. It is to solar astronomers what the *Lunar Orbiter Atlas of the Moon* is to lunar observers. Its glossy pages show sunspots the size of your fist and loops of gas forming magnetic arcades above the Sun's surface. Prominences rise themselves above the Sun using invisible magnetic forces and there are forests of strange structures seen close to the limb.

Some of the discoveries made from *Skylab* included coronal holes and X-ray bright points. Coronal holes were seen as dark regions in which the hot coronal material is very thin. X-ray bright points are small, compact, short-lived brightenings that are most easily seen inside the coronal holes themselves. The coronal holes were seen to maintain their shape through several solar rotations in spite of the variations in rotation rate of the solar surface. Significantly, the brightest coronal emissions came from above active sunspot regions.

Coronal holes helped solve a long-outstanding puzzle. Before *Skylab*, probes such as *Mariner 2* in 1962 detected fast streams in the solar wind, flowing not at 400 km per second but at 600 km per second or more. Curiously, they tended to recur at intervals of 27 days, which is the rotation period of the low-latitude Sun. What was more curious was that around the turn of the twentieth century, a series of magnetic storms were observed on Earth that tended to recur at 27-day intervals.

Skylab showed that both phenomena were associated not with sunspots but with the coronal holes, out of which seemed to come most of the solar wind. Apparently loops of magnetism above sunspots trap the plasma and hold it back, meaning that the solar wind can more easily escape away from sunspots; that is, from the holes in the corona. Above the poles of the Sun it was seen that there are two large, permanent coronal holes. A recent spaceprobe, *Ulysses*, found those regions filled with fast-moving solar wind, as expected.

Skylab's ATM also observed huge clouds of plasma rising every two days or so from the Sun. It was suspected that such bubbles – named coronal mass ejections – became interplanetary plasma clouds, some of which ignited magnetic storms when they reached Earth. The problem was that mass ejections are best observed when they rise above the limb of the Sun, and clouds moving that way will not hit Earth. However, some years later, at the end of 1983, the magnetospheric probe ISEE-3 (*International Sun–Earth Explorer 3*) pulled away from Earth towards comet Giacobini–Zinner and some time later was sufficiently far from Earth to intercept such mass ejections, showing that they were identical to plasma clouds near Earth.

More recently the Sun has been monitored in ultraviolet and X-rays by *Yohkoh*, a highly successful Japanese satellite. Its images give a clear and detailed view of coronal holes, coronal bright spots and mass ejections. A highly successful recorder of such ejections has been the LASCO instrument aboard the *Soho* (Solar and Heliospheric Observatory) satellite stationed at the Lagrangian L1 gravitational balance point on the sunward side of the Earth. Computer processing of the LASCO images made it possible for scientists to see ejections even when headed towards Earth.

A great deal has been learned about ejections since they were seen by *Skylab* in 1973, and it is now believed that much of the magnetic storm activity on Earth that was once credited to flares is actually associated with the ejections. Their energy apparently comes from the magnetic fields, their material from prominences and they need not come from sunspot regions.

Skylab was a remarkable success, but it was not a true space station in the sense that it was not designed to be resupplied with food and fuel. This was a pity, as it would be over 20 years before the United States had that type of facility again. *Skylab* was abandoned in February 1974 and re-entered the Earth's atmosphere in 1979.

Solar probes

Then came a series of probes designed to make a close approach to the Sun.

The two Helios probes were developed in Germany and launched by American Titan rockets from Cape Canaveral in Florida. They were designed to travel to within 50 million km of the Sun, closer than the innermost planet, Mercury's 56 million km. At that distance heat was a problem. Their surfaces were covered with special mirrors and solar cells and they turned every second to distribute the generated heat evenly. Despite this, the surface reached temperatures of several hundred degrees Celsius. *Helios 1* was launched on 10 December 1974, with *Helios 2* following on 15 January 1976. Contrary to expectations, the probes survived the hellish conditions without damage and still sent data to the earth even 10 years later. They are still orbiting the Sun, taking about 190 days to complete one orbit, a Helios 'year'.

Each Helios was equipped with 10 instruments that produced a vast amount of data. Personally I do not remember them fondly. When I was a junior astronomer at the Jodrell Bank Radio Observatory, being junior I often had to stay up all night to monitor observations. Radio astronomy can be carried out just as well during the night as by day, so naturally there is a desire to make it fit into convenient times, especially when you have someone at hand to do the inconvenient ones.

I had one observation run when two German astronomers had to take over the big dish at 3:30 a.m. to observe one of the Helios probes. So in they came at 3:00 and started to unplug my detectors, my time sources and frequency mixers, forcing an end to my observations for a while. When they went they left the observing room a mass of cables and it was up to me – only me – to reconnect them and restart my observations. A radio telescope is one of the most complicated beasts ever constructed by the hand of man. I had many unsuccessful attempts at getting the data flowing again. But when I did achieve it I felt I understood the system. When I handed in my doctoral thesis one of my colleagues read the second chapter and said, somewhat surprised, 'you definitely understand radio interferometers'. For a young radio astronomer that was praise indeed.

The *Solar Maximum Mission* (SMM or SolarMax) is a solar mission like no other. It was launched on 14 February 1980 and was a magnificent satellite carrying many scientific instruments to study the Sun through the high part of its 11-year solar cycle (the peak of the cycle was in 1979) and to provide new insights into the nature of solar flares and the Sun's total output, its 'solar constant'.

But disaster struck the project as a series of malfunctions cut short its work. In September 1980 the coronograph that produced images of the Sun's corona failed and a few weeks later the attitude control system, which enabled the satellite to point accurately, also failed. As stable pointing of the spacecraft was no longer possible, it was placed into so-called standby mode. It remained that way for more than three years.

However, NASA had a plan. SMM was designed in a modular fashion, which meant that its broken pieces could be removed and replaced with relative ease. So NASA decided that repairing SMM in orbit was a good demonstration of the benefits of manned spaceflight.

I recall the rescue mission well. Once the space shuttle *Challenger* was alongside, astronaut George Nelson was to use the powered backpack (technically the manned manoeuvring unit or MMU) to 'fly' between the shuttle and the slowly spinning SMM. He was to align himself with a docking clamp and fly into the SMM, making a connection with a docking attachment protruding from the front of the MMU. It sounded like science fiction and everyone knew it was a tall order. When the night of the procedure came we were all tense and I found myself watching the dramatic events from a television studio (for me that was a new experience).

Nelson reached SMM but could not dock. He used his gloves to try to grab the solar panels but made it spin worse. For a few moments things looked bleak. Eventually with the help of the shuttle's robot arm SMM was snagged, drawn in and latched down into *Challenger*'s cargo bay. Over the next few days it was repaired.

In June 1984, SMM resumed operations, capturing images of the corona during the daylight portion of each orbit. Its ACRIM instrument package monitored the total energy output of the Sun throughout the mission with great precision. It showed the expected dimming of the Sun when sunspots rotated into view and provided important information about the extent of this dimming. But it also showed that the Sun was brighter during the maximum of the sunspot cycle when more spots are observed on the Sun's surface, because although the sunspots themselves are dark and produce dimming, they are surrounded by faculae that are bright and more than offset the sunspot dimming.

However, observations were again interrupted in December 1986 when the coronagraph's dedicated tape recorder failed. Operation was restored in March 1987, with the data stored on the spacecraft's single remaining tape recorder. In this way observations continued without any major interruptions until the end of the mission. Atmospheric friction slowly caused the altitude of the SMM spacecraft's orbit to decline and on 17 November 1989 it lost altitude control and re-entered the Earth's atmosphere.

Other satellites had been operating over the period of solar cycles 21 and 22, including the US Air Force satellite *P78–1* and the Japanese *Hinotori* satellite. *P78–1* was launched in 1979 and carried the Solwind coronograph and various X-ray spectrometers, which observed the first broadening and shifts of X-ray spectral lines during the impulsive phase of a solar flare. It worked successfully until it was destroyed in an antisatellite missile test in 1985.

When it was first proposed, Ulysses was a spectacular mission involving two almost identical spacecraft launched into identical orbits above the Sun's poles, except that one would take the north pole while the other would take the south pole first, then both would make a far orbit of the Sun taking in a pass over the other pole. One spacecraft was to be built by the US and the other by the European Space Agency. But perhaps it was too good to be true. In the early 1980s the US had budget problems and pulled out, withdrawing too late for the remaining craft to incorporate the best choice of instruments. So the European probe went ahead a few years late and was launched, without a camera, in 1990.

The political setback aside, it proved to be an excellent mission which still continues. It is sampling the solar wind at solar latitudes unexplored by any other spacecraft. To reach above the poles of the Sun it carried a suite of instruments out to Jupiter, where that planet's gravity pulled the spacecraft into a trajectory that carried it over the Sun's south pole in the autumn of 1994 and its north pole in the summer of 1995. It has told scientists a great deal about the solar wind, especially how it changes from solar maximum to minimum.

But perhaps the most successful sun-monitoring spacecraft is *Soho*, which is helping us understand the interactions between the Sun and the Earth's environment better than ever before. Its scientific legacy may help scientists solve some of the most intractable riddles about the Sun, including the heating of the solar corona, the acceleration of the solar wind and the physical nature of the solar interior. *Soho* has, it is no exaggeration to say, provided astronomers with their first long-term, uninterrupted view of the Sun.

Although the scientific data prize goes to *Soho*, for an impression of the turbulent, shifting, appearing and disappearing, magnetic loops of hot gas, look no further than the wonderful images produced by the *TRACE* satellite. It has enabled solar physicists to study the connections between fine-scale magnetic fields and the associated plasma structures on the Sun's photosphere, the transition region, and its corona. Its images are beautiful and unsurpassed.

Active regions ripple with magnetic energy.

CHAPTER TWENTY-THREE

Sunjammer

There are plans, probes and problems that the current generation of solar astronomers bequeath to their successors because such is the timescale for many space missions, often more than a decade from conception to completion. It is rather like building a cathedral: everyone has their part. Science is like that. The vast majority of scientists add their own brick, placing it on others and allowing others to be placed on theirs.

All solar variability on human timescales is magnetic in origin and as we shall see has its root in the shifting rivers of plasma in the Sun's outer zone. As Hale suspected, the solar cycle is a magnetic cycle. Flares and mass ejections occur when magnetic fields are stressed beyond endurance. The structure of the corona and the solar wind is also determined by the Sun's magnetic field. One of the big mysteries of current solar physics is the heating of the Sun's corona and the acceleration of the solar wind. It is thought to have something to do with the interaction between small-scale magnetic structures. Answering those questions is the chief task of many of the missions to the Sun under construction or planning. One of them is the *Solar Dynamics Observatory*, which will look at how stored magnetic energy is unleashed.

There is also *Solar-B*, a Japanese Institute of Space and Astronautical Science mission proposed as a follow-on to the highly successful Japan/US/UK *Yohkoh* satellite. It consists of a suite of optical, ultraviolet and X-ray instruments that will look at the interaction between the Sun's magnetic field and its corona. It is scheduled for launch in the Autumn of 2005, when it will be placed in a Sun-synchronous orbit about the Earth that will keep its instruments in nearly continuous sunlight.

The *Stereo* mission is up next, scheduled to launch in February 2006. This two-year project will employ two nearly identical space-based observatories to provide the first ever 3-D stereoscopic images to study the nature of mass ejections. To obtain these unique views of the Sun, the twin observatories must be placed into a rather challenging orbit where they will be offset from one another. One will be placed 'ahead' of the Earth in its orbit and the other 'behind', by using a series of lunar swingbys to take them to the right position. Just as the slight offset between your eyes provides you with depth perception, this placement will allow the *Stereo* observatories to obtain 3-D images of the Sun. There is also a mission called *Solar Probe* that will pass closer to the Sun than Mercury so that it can take closeup measurements of the solar wind, in the hope that it will be able to glean clues about its acceleration. These missions will be joined by *Mercury Messenger*, due to be launched soon, which will measure the solar wind from orbit around the innermost planet, Mercury.

Many questions remain unanswered about the corona. Coronal holes persist at the solar poles through much of the solar cycle. During sunspot minimum, the polar holes are present and the solar wind is well organised. But the solar magnetic fields restructure themselves during the 11-year sunspot cycle, so that during sunspot maximum the polar holes are absent and the solar wind is mixed, with fast and slow winds seen at all latitudes. It is hoped that studying the restructuring of the solar magnetic field and its modifying of the corona and wind will give important information about the general sources of the fast and slow winds, tackling again the fundamental problems.

Sun sailing

The Sun could also give us the power for new types of missions that use its light in much the same way that sailors use the wind.

Imagine ... The space probe approaches its target just beyond the orbit of Mars. Approaching from its night side, it matches velocity and begins to scan the surface of the pock-marked, 10 km-wide rocky world. Data on surface topography and chemistry streams back to Earth. But as well as its scientific mission, the space probe has another function. After the asteroid has been mapped and scanned, a section of the probe detaches from the main craft and descends to the surface of the rocky world, where anchors attach it firmly as this tiny world's gravity is feeble. There it waits until the asteroid's orbit grazes the Sun.

A year later, harsh sunlight sears the sunward side of the asteroid. The passenger probe, protected by a sunshield, begins to stir as if coming out of hibernation. At the right time of the asteroid's spin, when the craft enters the Sun's shadow, it moves upwards but still keeps within its shadow. Then, like a butterfly beginning a terrestrial flight, it unfurls a thin metallic sheet a few kilometres across and emerges into the solar glare. Now brighter than any star, the metalic sheet strains and billows. The pressure from the Sun is filling it as the probe opens and takes flight. Like the seeds that leave a dandelion propelled by human breath, it starts to leave the close environs of the Sun and heads for the outer planets. It could be the Sun itself that provides us with a way to explore the outer solar system and beyond.

Light applies a slight pressure on any illuminated object. In 1924 space pioneers Fridrikh Tsander and Konstantin Tsiolkovsky noted that in the vacuum of space a large, thin sheet of reflective material could be a propulsion device requiring no propellant. Sunlight would literally push it along. For a highly reflective sail, the solar flux could produce a force of about 9 Newtons for every square kilometre (a Newton is the force needed to give a mass of 1 kg an acceleration of 1 metre per second, every second), which is quite reasonable since it is continuous.

In 1973 NASA sponsored a study for a solar sailing probe to intercept Halley's comet, pulling behind it a platform of scientific instruments. A team from the Jet Propulsion Laboratory designed a number of solar sails. The recommended JPL sailcraft had a central mast and booms that would spread and support a 2-micron-thick plastic sheet coated with aluminium, 850 metres square. Making one side of the sheet reflective and the other dark would set up a force imbalance that could be used for manoeuvring. It all weighed 5 tonnes, needed little highly advanced technology and had to be launched in 1985 to meet the comet. It was also decided that the Halley's comet interceptor should have a different form of propulsion, called a solar-powered electric propulsion system. Unfortunately both studies were soon rendered academic when the project was cancelled by Congress.

In 1979 the World Space Foundation, a non-profit-making organisation of space enthusiasts that included many JPL scientists, took up the idea again. In 1981 the world's first solar sail, a half-scale prototype, was exhibited. Two years later a full-size prototype was completed.

In the fall of 2004, deep beneath the surface of the Barents Sea, a Russian nuclear submarine may launch a single missile, left over from the old Soviet arsenal. Instead of a voyage of mass destruction, this rocket will be sent on a mission of hope for the future of mankind. It will launch on its way Cosmos 1, the first solar sail. The launch will be the culmination of the first international, privately funded space mission in history.

A spacecraft pulled along by a solar sail, a large metal sheet but microns thin that converts the momentum of sunlight into thrust, would be a very useful space vehicle. A 4 square km sail deployed from the space shuttle could take a shuttle cargo bay's worth of payload to Mars and then return to Earth for more. An ordinary rocket motor would get the equipment there sooner, but would need two to three times more fuel than cargo. When permanently manned Mars bases are established sometime later this century, six of these Sunjammers could form a bridgehead and lifeline for Mars colonists.

They could travel even further. The *Voyager 2* space probe currently leaving our solar system will take some 80,000 years to travel the 4.3 light years (about 25 million million miles) to the nearest star, though it is not heading in the right direction. A conventional solar sail could perhaps reduce this to about 15,000 years, still almost an eternity when compared with our individual lifespan. However, some scientists have suggested that we may be able to help Sunjammers along by firing lasers or microwave beams at them.

Studies indicate that a 3.6 km diameter solar sail trailing a 1 tonne probe could be powered by a 65,000 million Watt laser fired from Earth orbit. Such energies are far higher than any yet obtained, but could conceivably be developed during the next century. Although the laser light would initially be a tight beam, it would tend to spread out and diffuse over the vast distances of space. For this reason a large, so-called Fresnel lens 1000 kilometres wide located at a stationary point between Saturn and Uranus would be used to focus the laser light on the receding probe. Three years of such acceleration would give it a velocity 10% that of light, meaning that it would reach the nearest star system, the Alpha Centauri group, in 40 years. One could also make the solar sail out of an ultra-high-strength mesh and beam microwaves at it.

Such ideas, although fascinating, are still deep in the realm of speculation. Other propulsion technologies may be better at propelling the first exploration trips to the stars. But it is not too fanciful to think of future space mariners plying the trade routes between the planets on sails of gossamer that ride the sunlight.

Nearly 400 years ago astronomer Johannes Kepler observed comet tails blown by a solar breeze and suggested that vessels might likewise navigate through space using appropriately fashioned sails. It is now widely recognised that sunlight does indeed produce a force that moves comet tails, and a large, reflective sail could be a practical means of propelling a spacecraft. One concept explored by NASA was to develop an interstellar probe pushed along by sunlight reflected from an ultrathin sail. Nearly half a kilometre wide, the delicate solar sail would be unfurled in space. Continuous pressure from sunlight would ultimately accelerate the craft to speeds about five times higher than is possible with conventional rockets – without requiring any fuel! If launched in 2010, such a probe could overtake *Voyager 1*, the most distant spacecraft bound for interstellar space, in 2018, going as far in eight years as *Voyager 1* will have journeyed in 41 years.

Tennyson put it well:

> *For I dipt into the future, far as human eye could see,*
> *Saw the Vision of the world, and all the wonder that would be,*
> *Saw the heavens fill with commerce, argosies of magic sails,*
> *Pilots of the purple twilight, dropping down with costly bales.*

In many thousands of millions of years, our Sun will swell up and change from a yellow dwarf to a red giant, swallowing the Earth as it does so. For a while it will become much more luminous, so for a brief period Sunjammers would be more efficient. In that distant future age perhaps mankind, or whoever follows us, will use the light of our dying star to evacuate our Earth and propel us into the vast oceans of space, just as in a long-forgotten earlier age man once raised sails filled with wind to explore his home.

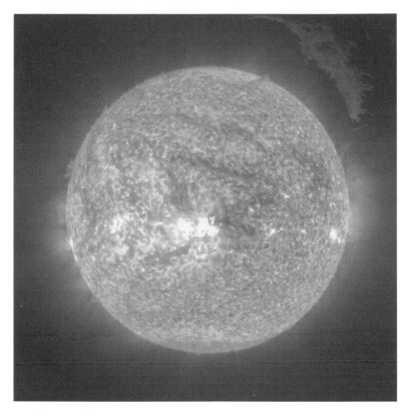

A huge prominence uses magnetic energy to rise above the Sun.

CHAPTER TWENTY-FOUR

Journey from the heart of the Sun

Circling the Sun, under the caverns of Mercury, Expedition Sundiver prepares for the most momentous voyage in history – a journey into the boiling inferno of the Sun. This is the setting for David Brin's 1980 novel about a mission to the Sun. It is a remarkable work, but such a journey is only fantasy as yet. Perhaps one day we will send unmanned probes to swoop down among the vast plasma arches that reach over the surface, sampling it, making coordinated measurements of plasma density, temperature, motions and magnetic fields. It would certainly go a long way to answer many questions. For example, Solar scientists believe that many of the outstanding mysteries of the Sun, such as the how the corona is heated or how the solar wind is accelerated, can be answered by the smallest-scale observations.

So now, with all the information we have gathered, all the scientists we have known and all the observations we have discussed, we are, at last, in a position to take a journey from the heart of the Sun.

We may never be able to go to the centre of the Sun, or at least we cannot imagine the technology that would allow us to visit the Sun's core, which, at a temperature of 15 million degrees, is too hot for atoms to exist. Here, in the innermost 10% of the Sun's mass, all is plasma – free protons, electrons, neutrons and helium nuclei at a density of 150 times that of water or about 10 times denser than the centre of the Earth. The particles are tearing around at hundreds of kilometres per second, but because of the density they are not able to travel very far, colliding many times a second.

In this chaos, two protons approach each other and there is a flash of light. The protons merge, forming an isotope of hydrogen – deuterium plus a positron (an antimatter version of an electron) and a neutrino. This is the start of the proton–proton chain. Then two deuterium nuclei fuse to produce helium-3 and a flash of high-energy radiation – a gamma ray. Then two helium-3 nuclei fuse, producing helium-4 and two protons. The net result of this is that hydrogen is converted to helium and energy. The helium nuclei that are newly made are lighter than the protons that make them up and the difference is converted into energy in the form of high-energy photons and neutrinos.

The Sun produces some 2 followed by 37 zeros of neutrinos and they all vanish instantly. It is a powerful reaction, but if you were to follow an individual proton situated at the core of the Sun you may have to wait a timescale of the order of a billion years for it to undergo a fusion encounter. It is rare for an individual proton to fuse, but then there are so many protons available. If taken outside of the star's core, a pinhead of this material would kill us at a distance of 160 km. So it is in this central zone that the Sun's energy is produced, the equivalent to 80 billion H-bombs going off every second as every second a million tonnes of mass is converted to energy. It has been like this for billions of years and will continue for billions more.

Gradually, the photons leak outwards, making their way through the dense plasma. Eventually we reach a point where the temperature is too low to sustain the proton–proton chain. The outside temperature is now about 9 million degrees and we have moved out of the region of energy generation to the radiative zone.

Radiative zone

This is an undramatic region where energy generated in the core down below is flowing outwards. Nothing changes here in this region of what physicists call hydrostatic equilibrium. The plasma at these depths has about the density of water and contains a huge amount of heat, but because it is surrounded on all sides by plasma of the same temperature it neither gains nor loses temperature. Energy is being moved out by radiative transfer. As we move outwards there is a very slight temperature gradient that drives the flow of energy outwards. The rate of energy flow is set by what the Sun produces at the core. This energy must escape otherwise the Sun will become unstable and explode, so the Sun adjusts the temperature fall between its layers so that it is just enough to drive the flow of the required energy through it.

The photons stagger outwards, performing what scientists call a random walk as they are emitted and absorbed. Slowly they travel outwards, but to get to the surface the photon must spend 200,000 years of being passed to and fro, each time losing a little energy as it does so. If the Sun's core were to switch off tomorrow, we would not notice any change in the Sun's output for a few hundred thousand years, although we would detect the decline in its neutrino flux. Because of this absorption and emission, the photon from the core starts off as a high-energy gamma ray and ends up as a photon of light. Although the total radiative energy does not change, its quality does and the photons become gradually downgraded in energy. It was discovered by studying the passage of soundwaves in the Sun's interior that the radiative zone rotates as a solid whole, a fact that has important consequences for the solar cycle. Curiously, there is some slight evidence that the core may be rotating slightly slower than the radiative zone above it.

Slowly the temperature falls as we move outwards until we have travelled 500,000 km. It is now 2 million degrees outside and the plasma is as dense as expanded polystyrene. We are approaching a fundamental boundary and change in the Sun. We are about to reach the convection zone. But first we come to the interface layer, which lies between the radiative zone and the convective zone. This thin layer has become more interesting in recent years as more details have been discovered about it.

The radiative zone rotates as one solid mass, but the pattern of differential rotation observed on the surface is consistent throughout the convection zone, which means that the radiative zone rubs against the convection zone at the interface layer. Here, the movement of electrically charged streams of gas and the interplay of rotation, convection and shear forces create magnetic fields that are the root of the solar cycle. The changes in fluid flow velocities across the layer can stretch magnetic fields of force and make them stronger. There also appear to be sudden changes in chemical composition across this layer.

Convection zone

The convection zone is 200,000 km thick and reaches almost to the surface. It is there because the plasma undergoes a change in its opacity. The photons that are the carriers of energy outwards are suddenly no longer able to travel very far and the flow of energy is impeded. The Sun has to do something, because as we have seen energy must escape. So it adjusts itself, and instead of transporting energy outwards in the form of radiation, it takes it outwards in the form of mass movement. In other words, it starts to convect. This is the region of huge streams of hot plasma, 100,000 km, wide that rise towards the surface at a few centimetres per second. When they reach the surface they give up their heat and then the now cooler plasma flows inwards, this time in narrower jets at a higher speed of a few metres per second.

A few thousand kilometres below the surface things change again. The hot updrafts fragment to a myriad of smaller convection cells, bubbles rising to the solar surface if you like. There is a hierarchy of structure here. Supergranules are much larger versions of granules (about 35,000 km across) and are best seen in measurements of the surface's Doppler shift, where light from material moving towards us is shifted to the blue while light from material moving away from us is shifted to the red. These features cover the entire Sun and the pattern is continually changing. Individual supergranules last for a day or two and have flow speeds of about 0.5 km per second. The fluid flows observed in supergranules carry bundles of magnetic fields generated in the convection zone to the edges of the cells, where they produce a so-called chromospheric network. The largest part of these are the wide supergranules, which are only a few thousand kilometres thick and last about a day, dissolving and reforming endlessly.

Closer to the surface are the mesogranules. They are about 5000 km in size and last a matter of hours. The granules are small (about 1000 km across) cellular features that cover the entire Sun, except for those areas covered by sunspots. These features are the tops of convection cells where hot plasma rises up from the interior in the bright areas, spreads out across the surface, cools and then sinks inwards along the dark lanes. Individual granules last for only about 20 minutes. The granulation pattern is continually evolving as old granules are pushed aside by newly emerging ones, a bubbling mass. The flow within the granules can reach supersonic speeds of more than 7 km per second, producing sonic booms and other noise that generates waves on the Sun's surface.

The science of helioseismology got its start in the early 1960s when CalTech physicist Robert Leighton discovered the five-minute oscillation on the surface of the Sun. Using the 60-foot solar tower, Leighton followed a suggestion made by George Hale many years earlier and looked at the effect of the Sun's magnetic field on the spectral lines from certain atoms. The Zeeman effect that Hale used to determine that sunspots were magnetic phenomena splits a spectral line into red and blue components, which indicate motion away from and towards an observer, respectively. Leighton and his colleagues, Robert Noyes and George Simon, took several spectroheliograms in the 'red' and 'blue' shifted light of specific solar lines from calcium. Analysing the data, they noticed something peculiar. Both large-scale and small-scale features on the solar surface were oscillating over a period of 296 seconds (or roughly 5 minutes) and at a speed of 500 metres a second. These oscillations quickly became known as solar quakes, and the study of helioseismology was born.

Later studies would show that these oscillations were due to sound waves trapped between regions of the solar interior of different densities. They, along with other oscillations subsequently detected, could be used to probe the inside of the Sun, and even make maps of what was going on over on its unseen far side. It also showed us that the Sun's convection zone was substantially deeper than our models of the Sun's interior had led us to believe.

There are many oscillations on the Sun that can be detected and the way they spread out allows the interior of the Sun to be studied, in just the same way that on Earth the study of shockwaves from earthquakes can be used, to study the internal structure of the Earth. There is now a network called Gong, the Global Oscillation Network Group, a series of telescopes scattered around the world so that the Sun is never left out of its sight, allowing long oscillations to be detected that would not be possible at one site because of the interference of night time. Many oscillations were seen for the first time in the long-duration observations possible from the South Pole.

But perhaps the strangest use of the Sun's surface waves is that they allow astronomers to see through the Sun. Teams in France and the US can now provide predictions of what the Sun could have in store for us by using two different ways of detecting activity on the Sun's far side, before it swings into view of the Earth. The Michelson Doppler Imager (MDI) onboard the *Soho* satellite looks at the sound waves on the Sun's surface to locate hidden far-side active regions. No longer will we be taken by surprise by highly active regions that suddenly come into view as the Sun rotates. The MDI detects changes in sound waves due to the presence of far-side sunspots and can analyse these changes to estimate the unseen sunspot's size and position. This is important, because as we have seen the Sun takes roughly four weeks to turn on its axis, but active regions can appear and grow in only a few days. For practical purposes astronomers have made the Sun transparent. The new long-range far-side forecasts will prove especially useful for scheduling manned spaceflights, during which astronauts might be exposed to dangerous radiation from solar flares.

Photosphere

We now arrive at the photosphere, which is the Sun's visible surface – although, since the Sun is a ball of gas, it does not have a conventional solid surface. The photosphere is actually a layer about 100 km thick and so very thin when compared to the rest of the Sun. Looking at the centre of the Sun we see somewhat hotter and brighter regions, but at the limb we see light from the upper, cooler regions. This explains the 'limb darkening' that appears as a darkening of the Sun approaching its edge. The 'surface' that we see is at a temperature of about 6000 degrees.

We have seen that the key to understanding so much on the Sun is magnetism. At the interface between the radiative zone and the convection zone, magnetic fields are produced by the complex motion of electrically charged gas. These magnetic fields form tubes of magnetism that rise upwards and loops of them break surface. In the regions of high magnetic field, the outward flow of energy is impeded and a cooler region ensues where the loops appear, often in pairs. Welcome to a sunspot.

An imaginary spacecraft braving the Sun's atmosphere and flying over a sunspot would provide its passengers with a spectacular sight. As you approach it, you would see there were changes in the granulation. They seem a little brighter and subtly distorted as if by some unseen force and the roughly circular granulation has been pushed into streaks and whirls. The dark mass – the size of a small planet – would loom on the horizon and as you reached its outer regions, the penumbra, you would notice strange light and dark filaments radiating outwards, constantly changing with the light and dark regions not always lining up with one another. Reaching the edge of the umbra – the central and darkest region – you would see the subdued convection in the spot as well as a zone of slightly brighter material at the centre. Islands of bright light would seem to dance at the sunspots very heart. You could look up and see high above you vast arcades of hot gas – a region that we shall shortly be visiting.

Sunspots tell us that the Sun rotates on its axis once in about 27 days and its axis of rotation is tilted about seven degrees with respect to the Earth's orbit, which means that we see more of the Sun's north pole in September each year and more of its south pole in March. But the Sun does not rotate rigidly: the equatorial regions rotate faster (about 24 days) than the polar regions (more than 30 days). Sunspots typically last for several days, although very large ones may live for several weeks. They are magnetic regions where the magnetic field strength is thousands of times stronger than the Earth's magnetic field. They often come in groups with two sets of spots. One set will have positive or north magnetic field while the other set will have negative or south magnetic field – an important clue about their origin. As we have seen, the magnetic field impedes the flow of energy from below so that they appear darker, though this darkness is a contrast effect. If you could see a sunspot on its own it would shine brighter than an arc lamp. The magnetic field is strongest in the umbra and weaker in the penumbra.

The most detailed pictures ever taken of the surface of the Sun were obtained recently by the new Swedish 1-metre solar telescope on the Canary Island of La Palma. The images show new solar features and hitherto unknown details in sunspots. A striking feature in the images of sunspots is the existence of dark cores within bright filaments. This is an unexpected discovery and astronomers are uncertain what it signifies, but they believe that something fundamental is taking place in the Sun's atmosphere on such scales.

To obtain the detailed view, the telescope's tube is evacuated and a mirror in the beam adjusts its shape 1000 times a second to counteract atmospheric blurring. Looking at the outskirts of sunspots, the researchers see that the penumbra appears to consist of thin, long filaments that have remained unresolved by solar telescopes until now. And scientists have even obtained a 3-D view of what it is like inside a sunspot by analysing sound waves travelling inside the Sun.

The picture that emerges is of fast-moving streams of hot, electrically charged gas converging into a gigantic vortex that reaches below the solar surface. Sunspots, it seems, are not static but consist of very strong, downward flows of plasma travelling towards the interior of the Sun at speeds of about 4800 km per hour. To map the interior of a sunspot, scientists used *Soho*'s Michelson Doppler Imager again, dissecting a single large sunspot that was visible on 18 June 1998. By measuring the speed of solar sound waves generated that day, the researchers were able to produce 3-D maps of a region extending about 16,000 km below it.

Analysis revealed that sound waves travel about 10% slower at the surface where temperatures are lower, and maintain this relatively slow pace as they begin moving towards the interior of the Sun. When the sound waves reach a point about 4800 km below the surface, however, their speed increases significantly, indicating that the roots of a sunspot are hotter than their surroundings and thus implying that sunspots are cool only to depths of about 4800 km – a relatively shallow layer considering that it is about 692,000 km from the surface to the centre of the Sun. One of the striking features of these observations is just how shallow a sunspot is.

Most easily seen near the limb but present everywhere are the faculae – bright magnetic areas where the magnetic field is concentrated in much smaller bundles than in sunspots. While sunspots tend to make the Sun look darker, faculae make it look brighter. In fact, during a sunspot cycle the faculae dominate and make the Sun appear slightly (about 0.1%) brighter at sunspot maximum that at sunspot minimum.

Chromosphere

Above the photosphere we reach the chromosphere, an irregular layer where the temperature rises from 6000 degrees Celsius to about 20,000 degrees. At these higher temperatures hydrogen gives off a reddish colour that is especially interesting to see in the prominences that project above the limb of the Sun, so often seen during total solar eclipses. In the colour of a particular element it is possible to make out a kind of chromospheric network of magnetic field elements, bright regions around sunspots, dark filaments and prominences. The network outlines the supergranule cells and is due to the presence of bundles of magnetic field lines that are concentrated there by the fluid motions in the supergranules.

Look closely at sunspots and you may see bright patches surrounding them. These are called plage, the French word for beach, and they too are best seen in the red light of hydrogen. They are associated with concentrations of magnetic fields and clearly form a part of the network. Magnificent to see are the so-called filaments, which are dense, somewhat cooler huge clouds of material that are suspended above the solar surface by magnetic forces. Prominences and filaments are actually the same things. Then there are the spicules, small, jet-like eruptions seen throughout the chromospheric network. They last but a few minutes but in the process eject material off the surface and outwards into the hot corona at speeds of 20 to 30 km per second.

Transition region

Moving upwards, we next come to the transition region, which is a thin and very irregular layer of the Sun's atmosphere that separates the hot corona from the cooler chromosphere. Here the temperature changes rapidly from 20,000 degrees Celsius to 1,000,000 degrees. Much of this region is only really observable from space. The transition region has been studied from space using instruments on several spacecraft, including the *Soho* satellite and the Transition Region and Coronal Explorer (TRACE) mission that is now actively acquiring data on the structure and dynamics of the region.

Corona

Finally we come to the Corona – the Sun's outer atmosphere. As we have seen, it is visible during total eclipses of the Sun as a white light surrounding the Sun, and outside eclipses using a coronograph, but it has only been known since about 1940 that it is so hot. Inside the corona are a variety of features including streamers, plumes and loops. Here at temperatures greater than a million degrees both hydrogen and helium are completely stripped of their electrons. Even minor elements like carbon, nitrogen and oxygen are reduced to bare nuclei. Only the heavier trace elements like iron and calcium are able to retain a few of their electrons in this intense heat. It is emission from these highly ionised elements that produces the spectral emission lines that were so mysterious to early astronomers.

From the upper corona the solar wind streams off the Sun in all directions at speeds of about 400 km per second, because the corona is so hot that the Sun's gravity cannot hold onto it. Although we understand why this happens, we do not understand the details about how and where the coronal gases are accelerated to these high velocities. This question is related to coronal heating. How the 1,000,000 degree corona, which sits above a 6000 degree surface, gets so hot is a mystery. Look around in the corona and you may see polar plumes, which are long, thin streamers that project outwards from the Sun's north and south poles. The footpoints of these features are associated with small magnetic regions on the solar surface.

Coronal loops are found around sunspots and in active regions. These structures are associated with the closed magnetic field lines that connect magnetic regions on the solar surface. Many coronal loops last for days or weeks. Some loops, however, are associated with solar flares and are visible for much shorter periods. These loops contain denser material than their surroundings. The three-dimensional structure and the dynamics of these loops are an area of active research.

As *Skylab* first showed, coronal holes are regions where the corona is dark. The high-speed solar wind is known to originate in coronal holes. The solar wind is not uniform. Although it is always directed away from the Sun, it changes speed and sometimes carries with it magnetic clouds. The solar wind speed is high, 800 km per second, over coronal holes and low, 300 km per second, over streamers. These high- and low-speed streams interact with each other and alternately reach the Earth as the Sun rotates, buffeting the Earth's magnetic field and sometimes producing magnetic storms.

Spiralling out from the sun like a gigantic spiral lawn sprinkler are so-called corotating interactive regions, which are places within the solar wind where streams of material moving at different speeds collide and interact with each other. As the Sun rotates these various streams produce a spiral pattern in the solar wind. But if a slow-moving stream is followed by a fast-moving stream, the faster-moving material will catch up with the slower material and plough into it. This interaction produces shockwaves that can accelerate particles to very high speeds. Curiously, the composition of the solar wind is not the same as the Sun's surface and it shows variations that are associated with solar activity and solar features.

But let us travel back down from the corona – a solar flare is about to take place.

Solar flares

Arcades of magnetic flux tubes arch over a sunspot region. As the coil emerges from beneath the photosphere, it becomes filled with tenuous, hot coronal gas, lighting up its outline. The coil rises, but movement at its footpoints shows that they are twisting. This differential movement causes the magnetic field to become tense and strained. It becomes distorted and amplified, until eventually it breaks and reconfigures itself into a simpler configuration that takes less energy to maintain. But the energy it once had has to go somewhere and it violently, explosively, heats the surrounding gas. The plasma becomes superhot, shockwaves ripple out and beams of accelerated electrons can fly in any direction. If they strike the surface they can cause a hot spot that contributes to the light of the flare. Electrons also spiral through the magnetic fields giving off radio waves, either as static or in falling frequencies as they slide along the field lines giving up their energy.

In a matter of just a few minutes, flares heat material to many millions of degrees and release as much energy as a billion megatons of TNT. They occur near sunspots, usually along the magnetically neutral dividing line between areas of oppositely directed magnetic fields. They release gamma rays and X-rays, protons and electrons. The biggest flares are X-Class flares. M-Class flares have a tenth the energy and C-Class flares have a tenth of the X-ray flux seen in M-Class flares. In the hours following a solar flare you can see a series of loops above the surface of the Sun.

Looking closely you can see how material 'condenses' out of the Sun's hot corona in the tops of these loops and then flows down the legs of the loops onto the surface. Within the magnetic confines of these loops the material is somewhat isolated from the million-degree corona and can cool to much lower temperatures.

All the time vast mass ejections are preparing to lift off the Sun. They are huge bubbles of gas threaded with magnetic fields that are ejected over the course of several hours. Although the Sun's corona has been observed during total eclipses of the Sun for thousands of years, the existence of coronal mass ejections was not realised until the space age. The earliest evidence of these dynamic events came from observations made with a corona-graph on the seventh Orbiting Solar Observatory from 1971 to

1973, and of course *Skylab* saw them. But these seemingly chaotic expulsions have a hidden order and play an important role. Coronal mass ejections are often associated with solar flares and prominence eruptions, but they can also occur on their own. The frequency of ejections varies with the sunspot cycle. At solar minimum we observe about one ejection a week. Near solar maximum we observe an average of two to three ejections per day.

Eight years of ejection observations by *Soho* shows that they are removing the Sun's old magnetic field bit by bit, first from one pole and the equator, and then the other pole. In a way, the Sun is shedding its old magnetic field like a snake. More than 1000 coronal mass ejections, each carrying billions of tonnes of gas from the polar regions, are needed to clear the old magnetism away. But when it is all over the Sun's magnetic stripes are running in the opposite direction.

We are now moving away from the Sun, leaving behind its fiery, turbulent surface. We have the solar wind at our backs as we travel out into the realm of the planets. Our journey from the centre of the Sun to its limits is far from over, however.

Magnetic fields

We have learned so much over the centuries about our local star. At first we measured its motion, then we analysed its light, then its surface and finally what lies beneath the surface was analysed by sound waves and neutrinos. So much has been discovered, yet so much remains to be discovered.

As we have seen, it is believed that the solar cycle is a magnetic cycle caused by the circulating of solar magnetic fields. The Earth's magnetic field is, by comparison, relatively simple to explain by the motion of fluidic iron in the outer regions of our planet's core. Currents in the iron constitute electrical circuits that generate magnetic fields around them – magnetic fields that reach high beyond the confines of the planet and out into space.

Occasionally the Earth's magnetic field vanishes, or at least declines to a low level, and then re-emerges with the opposite polarity. This we can measure by the opposite magnetic stripes seen in sedimentary rocks laid down over the transition. We do not know what causes the Earth's magnetic field to flip this way, but we do suspect that a flip could take place relatively soon. There can be periods without reversals for many millions of years and then we can have four or five reversals within one million years. What we do know, however, is that the current decay in the Earth's field started 2000 years ago.

Over the last century and a half, since monitoring began, scientists have measured a 10% decline in the Earth's dipole and at the current rate of decline it would take 1500 to 2000 years to disappear. A particular weakness in the field has been observed off the coast of Brazil in the so-called Southern Atlantic Anomaly. Here, eccentricities in the Earth's core have caused a dip in the field, leaving it 30% weaker than elsewhere. The extra dose of radiation creates electronic glitches in satellites and spacecraft that fly through it. Even the Hubble telescope has been affected.

In the past it seems that magnetic reversals were always preceded by weakened magnetic fields, but not all weakened fields brought on a flipflop. It is possible that the Earth's invisible shield could also grow back in strength and that sometime, maybe 10,000 years from now, the dipole will decay again and that will lead to a reversal.

Some scientists have speculated that when the Earth's magnetic field declines in advance of a reversal, the flux of cosmic rays from deep space that can reach the Earth increases. This, they argue, causes an increase in the background radiation, which leads to an increase in the mutation rate for living things on Earth, which in turn means a spur in evolution. It is an interesting idea, but it now seems that while the Earth is without its magnetic sheath the Sun will still keep us protected. When seafloor sediments from the mid-points of the spreading oceans are examined for their magnetic signatures, we can see evidence of 171 reversals in the Earth's magnetic field in the past

76 million years. We are due for another one, but we may take comfort that even if our compasses point neither north nor south, the Earth's creatures have survived reversals intact many times before and so will we, thanks to the Sun.

We must now delve back inside the Sun, to the interface region between the radiative zone and the convection zone. The origin of the Sun's 11-year sunspot cycle lies there.

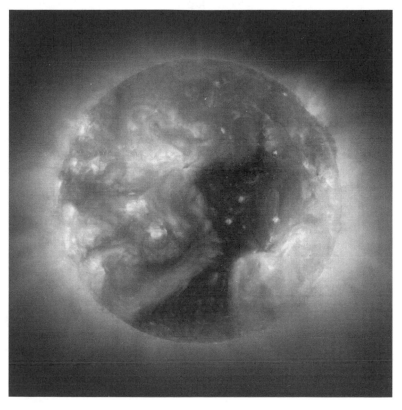

A dark 'coronal hole' spreads over the Sun's surface, allowing the solar wind to escape.

CHAPTER TWENTY-FIVE

Alpha Omega

A stronomers believe that the patterns of sunspots, the looping shape of solar flares, the solar wind, the mysterious sunspot cycle and the continual dispersal of plasma clouds into space are all controlled by magnetic energy generated in the Sun's interior by a dynamo.

A dynamo is something that converts mechanical energy into magnetic or electrical energy. There are many examples around us. It is a device that can be attached to a bicycle wheel, converting the rotational motion of a small wheel, driven by contact with the bicycle's wheel, inside which moving magnets produce an electric current that powers the headlamp. In a hydroelectric power plant, water behind a dam is channelled downwards, converting gravitational energy into kinetic energy that then drives a generator like the one on the bicycle to produce electric current.

As we have seen, the solar interior is in a state called a plasma, so there is an abundance of free-flowing electric charges. In such conditions scientists say that the magnetic field is 'frozen' into the plasma, because any motion of the plasma will carry the magnetic field along with it. What happens is that the motion of the plasma carrying its magnetism distorts pre-existing magnetic fields in such a way as to produce electric currents that will themselves induce a secondary magnetic field, which is in turn distorted by plasma motions that produce new electric currents, and so on in an interplay of movement, magnetism and electricity. It is this feedback that is at the heart of the Sun's rhythm.

Magnetic fields are like rubber bands. They consist of continuous lines of force that have tension and pressure. Like rubber bands, magnetic fields can be strengthened by stretching them, twisting them and folding them back on themselves. This stretching, twisting and folding is done by the plasma flows within the Sun during a dance of rotation and convection.

The observed surface rotation, which astronomer Richard Carrington so exquisitely mapped, is faster at the equator than at the poles and it is a pattern that *Soho* observations have shown carries on beneath the surface down to the base of the convection zone. Below that the rotation becomes like a solid body at a rate equal to the surface mid-latitude. This means that there are regions of shear where, just like the stretching rubber band, magnetic fields are dragged along, stretched and compressed by the plasma's complicated contortions.

The magnetic fields generated within the Sun are stretched out and wound around the Sun by its differential rotation – called the omega effect. The Sun's differential rotation can take a north–south-oriented magnetic field line and wrap it once around the Sun in about eight months. Twisting of the magnetic field lines is caused by the effects of the Sun's rotation, called the alpha effect. Early models of the Sun's dynamo assumed that the twisting is produced by the effects of the Sun's rotation on very large convective flows that carry heat to the Sun's surface. But one problem with that assumption is that the expected twisting is far too much and it produces magnetic cycles that are only a couple of years in duration. More recent dynamo models assume that the twisting is due to the effect of the Sun's rotation on the rising magnetic 'tubes'. The twist produced by the alpha effect makes the magnetic field reverse from one sunspot cycle to the next, as first observed by Edwin Hale. In essence, magnetic fields, twisted and distorted, become complex and then readjust themselves to be simple again in an endless cycle – the solar cycle.

Further evidence that the root of the sunspot cycles is in the shear regions at the base of the convective zone comes from *Soho* observations. A team of Stanford scientists has narrowed the search for this region to a layer 61,000 km thick and centred at a depth of about 217,000 km below the Sun's surface, a region where there is the expected high level of turbulence and shear flows caused by changes in rotation rate.

Before *Soho*'s observations, ground-based studies indicated that the shear layer in this region was fairly broad, and overlapped significantly with the convection zone where the dynamo could not exist. Because of this, some doubted the dynamo theory altogether. *Soho* solved this problem, showing that the shear layer is sharper and does not extend into the convection zone as feared. Also, observations indicate that sound waves speed up more than expected in this region, suggesting that the turbulence and mixing associated with a dynamo are present.

A slow motion of plasma observed on the surface from the equator to the poles is now considered to be an important component of the solar dynamo. Previously, scientists believed that this equator-ward drift was a wave-like process involving magnetic forces. However, new evidence suggests that this drift is produced by a giant circulation system in which the compressed gases, 200,000 km below the Sun's surface, move from the Sun's poles to its equator at about 5 km per hour – a leisurely walking pace. The gases then rise near the equator and turn back towards the poles, travelling in the surface layers where the gas is less compressed – moving at a faster rate of approximately 30 to 60 km per hour. Recent progress in theoretical modelling of the sunspot cycle has emphasised these vast subsurface currents that are larger than our Earth. The speed of this circulation system, called the meridional circulation, changes slightly from one sunspot cycle to the next. The circulation is faster in cycles shorter than the average 11-year period and slower in cycles longer than the average period. This is a strong indication that this circulation acts like an internal clock that sets the period of the sunspot cycle.

The next sunspot cycle is referred to as cycle 24 because of the numbering system that dates back to the eighteenth century. We do not know what it will bring, but we have some clues based on observations of the meridional flow. Predicting features of the solar cycle may help us anticipate sunspots and solar storms and mitigate all the problems we have seen that they can cause.

The world-sized rivers of plasma that circulate between the Sun's equator and its poles over a number of years seem to be key to understanding and predicting the solar cycle. Forecasters believe the next solar cycle will begin in 2007 to 2008 if this plasma circulation, which has slowed down during the present solar cycle, continues to decelerate. That would mean that cycle 24 would begin about a half-year later than if it followed the standard 11-year span. A computer model, known as the Predictive Flux-transport Dynamo Model, successfully accounts for the 11-year duration of the solar cycle as well as such mysterious events as the reversal of the Sun's magnetic north and south poles that occurs towards the end of each cycle. It focuses on the meridional flow pattern of plasma moving between the equator and the poles over a period of about 17 to 22 years.

The circulation appears to influence the strength of future cycles, as seen in the number and sizes of the sunspots produced, not in the cycle immediately following but rather in a two-cycle or 22-year time lag. When the flow is fast, it concentrates the magnetic field at the Sun's poles. These stronger fields are then transported downwards into the solar interior, where they are further compressed and amplified to become the intense magnetic fields that form sunspots years later.

This flow is believed to transport the magnetic 'imprints' of sunspots that occurred over the previous two sunspot cycles. Indeed, how the magnetic flux from decayed sunspots is recycled back into the Sun is an important current area of research. By analysing these past solar cycles, scientists hope eventually to forecast sunspot activity about two solar cycles, or 22 years, into the future.

The Sun is now in the declining phase of the current sunspot cycle that peaked in 2000 and 2001. Because the circulation flow was faster than average during the previous cycle, astronomers believe that the next cycle will be a strong one, peaking in the years 2010 and 2011.

It seems that the sunspot process begins with tightly concentrated magnetic field lines in the convection zone that inhibit the flow of heat causing a slightly cooler region where they break surface. The field lines rise to the surface at low latitudes and form bipolar sunspots. When these sunspots decay, they imprint the moving plasma with a magnetic signature. As the plasma nears the poles, it sinks and starts returning towards the equator. The increasingly concentrated fields become stretched and twisted by the internal rotation of the Sun as they near the equator, gradually becoming less stable than the surrounding plasma. This eventually causes the coiled-up magnetic field lines to rise up, tear through the Sun's surface and create new sunspots.

This idea would explain many things. Since the plasma flows towards the equator, the theory explains why sunspots appear mostly in the Sun's mid-latitudes early in the solar cycle and then gradually become more common near the equator. Sunspots also become increasingly powerful with the progress of the solar cycle, because the continuous shearing of the imprints of the magnetic fields by the denser plasma beneath the surface of the Sun increases the strength of the spot-producing magnetic fields.

Stars like the Sun

This solar cycle work may also have implications for understanding stars like our Sun. It seems that the faster stars like our sun rotate, the more disturbances they experience, possibly indicating that their dynamo operates in a higher gear, transporting spots more quickly and creating bigger, more active starspots. It would be fascinating to observe such stars, with their large spots, huge flares and rapid solar cycles. I wonder what their effects would be on any form of life living on a planet orbiting them.

Which leads us to the important question: can we detect solar-type cycles occurring on other stars? Only if we can detect stellar activity cycles occurring in similar and in different types of stars can we truly put our Sun into its stellar perspective. Is our Sun's cycle unusual, is it fast or slow, and do other stars exhibit Maunder-like minima? Indeed, can we look at other solar-type stars and gain some clues as to how often Maunder-like minima occur in stars like our Sun?

The study of stellar activity cycles owes a lot to a Californian, Olin Wilson. He was educated at the University of California at Berkeley and the California Institute of Technology, where he earned his doctorate. He spent his entire career at the Mount Wilson Observatory, where he made spectroscopic studies of stellar chromospheres and stellar activity cycles. From March 1966 he monitored 139 stars, 'for the purpose of initiating a search for stellar analogues of the solar cycle'. He examined the spectral lines of ionised calcium, which indicates enhanced surface activity, and he showed that other stars do have regular cycles of activity, with some of them stuck in Maunder-like minima. From his work and its follow-ups, it was estimated that stars like our Sun spend about 10% of their time in a Maunder minimum. Or so it was thought.

The latest research shows that nearly all the supposedly Sun-like stars displaying minimal activity are, in fact, much brighter than and significantly different from the Sun and therefore not exact examples of Maunder minima in a Sun-like star. The findings throw into question all studies using these stars to make inferences about the Sun's own activity and future minima. The vast majority of stars identified as Maunder minimum stars are well above the main sequence, which means that they are not Sun-like at all, being either evolved stars or stars rich in metals like iron and nickel. To date, astronomers have not found a solar-type star that is unambiguously in a Maunder minimum state.

As well as Olin Wilson's survey, which was ably continued by Sallie Balliunas, there are other observational programmes designed to detect solar-type cycles in other stars. The Lowell programme, which began in 1984, was a photometric survey of 35 targets, undertaken to address our lack of understanding of brightness variations in Sun-like stars. It has been superseded by high-precision nightly observations of 350 Sun-like stars made with 0.8-metre automated photometric telescopes in Arizona. Long-term observations of G dwarfs in the star cluster M67 are also being carried out.

Added to this are observations of Sun-like stars conducted from space. These started with the *Nimbus 7* satellite, launched in 1978, and the *Solar Maximum Mission*, which began an unbroken series of stellar data that continues today with other satellites. The *Sorce* spacecraft launched on 25 January 2003 continues this work with its precise measurements of solar radiation. *Sorce* – the Solar Radiation and Climate Experiment – provides measurements of incoming X-ray, ultraviolet, visible, near-infrared and total solar radiation. Data obtained by *Sorce* has become crucial to explain and predict the effect of the Sun's radiation on the Earth's atmosphere and climate.

The key to looking at other stars to obtain specific information is in finding stars that are exactly like our Sun. It is not as easy as it might be imagined, since despite the fact that our Sun is an average star and there are billions of them out there, getting one in the sights of our telescopes is another matter. There have been many systematic searches for solar analogues and some of the searches have been better than others. At present we only have one star, 18 Scorpii, that is a good enough match to be called the Sun's twin.

A plume of plasma rises above the Sun's surface.

CHAPTER TWENTY-SIX

Among the stars

*T*o place our Sun in its rightful place among the stars, let us first look at its neighbours. The Sun's nearest neighbour is really a system of three stars orbiting one another. The brightest of these is Alpha Centauri A, a star very similar to our yellow Sun. Alpha Centauri B is a bit smaller and its dimmer light has an orange hue because the temperature of its surface is cooler – about 4800 degrees Celsius compared to the 5800 degrees of our Sun. The colour of a star tells us its temperature: the coolest stars are red and hotter ones are orange, yellow and bluish-white.

The two main stars in the Alpha Centauri system take about 80 years to orbit each other. They are quite widely separated, about 20 times as far from each other as Earth is from the Sun or the distance from the Sun to the planet Uranus. The third star in the system is Alpha Centauri C or Proxima Centauri because it is the closest to us. It is a much more representative member of the stellar fraternity, being a dim, red (therefore cool), small star. It is a great distance from the chief pair, some 300 times the Sun–Pluto distance. If our Sun had a companion star like Alpha Centauri C the same distance away, it would look like a very ordinary star in the night sky. It would be visible to the unaided eye, but it would be undistinguished and much dimmer than hundreds of other stars.

Also in our cosmic neighbourhood is Barnard's Star, named after Edward Emerson Barnard, who lived about a century ago and was said to be one of the keenest-eyed astronomers of all. It is a modest little star, just six light years away in the direction of the constellation Ophiuchus. It is the closest star that can be studied from the northern hemisphere with the aid of telescopes, but few astronomers do. It bears a strong resemblance to Proxima Centauri and is what is termed a red dwarf, the most common type of star in the galaxy.

Red dwarfs have about 10–30% of the mass of our Sun. They run through their nuclear fuel at a slow rate and, at their steady pace, will have lifetimes of tens of billions of years. But they are far from uninteresting and their study does help us understand our own star. The outer layer of our Sun is a region of convection, but these stars have much more vigorous and deeper convection zones. In fact, some of them may be completely convective. This results in the generation of strong magnetic fields and, when those fields rise above the star's leeringly red surface, dramatic explosions occur.

The stellar flares from these dwarf stars are much more energetic than the flares we see on our own Sun. They were first detected as stars that seem to flare suddenly for a few minutes and they were called, unsurprisingly, flare stars. Radio waves have also been detected from these titanic stellar flares. This was first done by Sir Bernard Lovell in 1959 with the then new large radio telescope at Jodrell Bank. Many years later, a young postgraduate student spent many night-time hours at the controls of the same radio telescope, using new techniques to detect stellar flares from a handful of red dwarf stars in nearby space. My thesis on this work is in the library at Jodrell Bank.

One of the stars we observed was a puzzle. When we looked at it one year we detected a number of flares, but a few years later it had gone quiet. I remember writing in my notebook, 'Is this a stellar activity cycle like our Sun's 11-year cycle?' Maybe.

Barnard's Star is racing through space and its apparent motion across the sky is one of the fastest known, but it is too small and faint to affect the shape of any constellation. The constellations of the sky seem to be unchanging, and to all intents and purposes over a human lifetime or two they are. But the stars do slowly change their positions over the centuries as the Sun and the other stars move around the centre of our galaxy, taking 200 million years for an orbit. It is a slow process – the constellations of 10,000 years ago are recognisable, but go back a million years and today's stargazer would be lost, although with the joy of a fresh sky to chart. Barnard's Star moves across the sky at a rate of about half a degree (the size of the Moon's diameter) every 175 years. It is getting closer and will pass by us at a distance of just under four light years (closer than Proxima Centauri) in about AD 11,800.

For many years some astronomers thought that Barnard's Star had a planet in orbit around it. Observations suggested that in its motion across the sky it was 'wobbling' slightly around a straight-line path. It is possible that this wobble was due to the gravity of a large planet or planets. But there are problems with the data, as the wobble is on the verges of detectability. In the past ten years or so we have discovered that planets orbiting stars in the solar neighbourhood are commonplace, and therefore the universe must be full of planets.

There is another red dwarf near us that has become famous because of the television series *Star Trek*. It is Wolf 359, the scene of a spectacular battle between the United Federation of Planets and an invading force from the Borg – a collective society where individuals do not exist – intent on 'assimilating' Earth. It is in the constellation Leo and is the faintest star among our neighbours, indeed one of the dimmest stars known. If the Sun were to be replaced by Wolf 359 there would be no proper daylight on Earth. Its light would be only 10 times as bright as full moonlight.

There are other red dwarfs close by. Lalande 21185, in the constellation of the Great Bear, is one. Then there is UV Ceti, a pair of red dwarfs and the prototype for the whole class of flare stars that includes Proxima Centauri and Wolf 359. The UV Ceti pair are about six times as far apart as Earth is from the Sun and they take about 25 years to orbit around one another. Their total mass is only about 30% of the Sun's.

The brightest jewel in the vicinity of the sun is Sirius or the 'Dog Star', in the constellation Canis Major. Sirius is a double star. Sirius A is a blue-white star about twice as big as our Sun. Its surface is at a temperature of 10,000 degrees. Its small companion, Sirius B, is our nearest example of a white dwarf star: the dense, collapsed remains of a star that long ago ran out of nuclear fuel. It is only about the size of the Earth but has about the mass of our Sun, making its material so compressed that a cup full of it would have the mass of a jumbo jet. You would weigh 100 times more than you do on Earth if you could stand on its surface. These two radically different stars, orbit around one another in about 50 years, with an average distance between them of roughly 20 times the Earth–Sun distance. The last of the stars we know to be closer than 10 light years is Ross 154 and it is again a red dwarf.

In 1783, musician turned astronomer William Herschel, published his observations leading to the discovery of the solar motion. He determined that our solar system is moving between its neighbouring stars in the direction of the star Lambda Herculis or Maasym, which is Arabic for wrist. Herschel used the term solar apex for this direction. The sky's brightest star, Sirius, is the solar 'antapex' – the point that our Sun appears to be travelling from.

This is the direction of our Sun's motion in its orbit around the centre of our Milky Way galaxy. All of the 100,000 or so stars in our galaxy orbit the centre. Those close to the core do so swiftly, whereas the Sun, being about 24,000 light years from the centre, moves in its orbit at a speed of about 220 km per second, completing an orbit in about 230 million years; so it has been around the galaxy many times, approximately 18 times. But as well as going around the galaxy it is also going up and down, rising above and dipping below the plane of the galaxy in an oscillatory fashion. The period of oscillation is about 70 million years. This means that we pass through the galactic midplane about every 35 million years, which some people have compared with the interval between mass extinctions on Earth. It is true that the number of cosmic rays striking the Earth will increase during the 100,000 years we are closer to the galactic plane, and perhaps they will have an effect on the Earth's cloudiness and thus its climate.

Our galaxy has a series of spiral arms and our Sun is currently located in the small spiral arm that we call the Orion arm (or local arm), which is a connection between the two nearest major spiral arms – the Sagittarius and Perseus arms. We pass through a major spiral arm about every 100 million years, taking about 10 million years to travel through. During the transit, there would be a higher rate of 'nearby' supernova whose intense radiation, even from tens of light years away, could alter the climate of the Earth.

Classifying stars

Our Sun, then, is the second brightest of the very closest stars sharing their orbit around the galaxy's centre. But we have selected too small a volume for us to be able to place the Sun accurately in its cosmic context. If we expand further but a little way into our galaxy we see brighter stars and larger stars as well. To understand where our Sun fits in among them, we have to consider the lives of the stars, and we can do that with the aid of a remarkable diagram.

It is obvious to any stargazer that stars have different colours. The white of Sirius is in stark contrast to the red of Betelgeuse in Orion. After the invention of the spectroscope, stellar spectra were obtained and a process of classification began. The system we use today was devised by Annie Jump Cannon (1863–1941), who will be forever associated with the phrase, 'Oh, Be A Fine Girl – Kiss Me!' This is the way that generations of astronomers have remembered the spectral classifications of stars.

Annie Jump Cannon was the eldest of three daughters of Senator Wilson Cannon, a Delaware shipbuilder, and his second wife Mary Jump, who taught her daughter the constellations (one of the duties and joy of a parent) and instilled in her a love of the night sky. She studied physics and astronomy at college, but on her graduation in 1884 she returned to Delaware for a decade.

In 1896, she became a member of the group later called Pickering's Women, who were hired by Harvard College Observatory director Edward Pickering to do the routine calculations that astronomers required. Pickering said that 'the first step is to accumulate the facts' and his goal was to classify the stars according to their spectral appearance. The analysis was begun in 1886 by Nettie Farrar, who left after a few months to be married. Her place was taken by Williamina Fleming, who scrutinised the spectra of more than 10,000 stars and developed a classification system containing 22 categories. The work was carried on by Antonia Maury, who developed her own system that was according to some accounts rather cumbersome.

It was left to Annie Jump Cannon to continue the task and she paid particular attention to the spectra of bright southern-hemisphere stars. She amalgamated Fleming's and Maury's ideas into the spectral classes O, B, A, F, G, K and M. Like that of Cecelia Payne, Jump's career spanned more than 40 years, during which women in science won grudging acceptance. She received many honours such as an honorary doctorate from Oxford, but it was only in 1938, two years before her retirement, that she obtained a regular Harvard appointment.

W stars are white or bluish and have surface temperatures of greater than 40,000 degrees. O stars are white and slightly more bluish, as are B stars. Type A are white again, F and G becoming more yellow, K orange and M red. Sirius is a type A, red dwarfs are type M, Alpha Centauri B is type K, and our sun is a type G star.

The next step forward in understanding stars was taken in the early twentieth century by the work of the astronomers Ejnar Hertzsprung and our old friend Henry Norris Russell. It led to the construction of what is now known as Hertzprung–Russell (H–R) Diagrams. In an H–R diagram a star's intrinsic brightness is plotted against its surface temperature (or spectral type). When this is done the result is extraordinary.

Most stars fall on a narrow band known as the main sequence that contains stars ranging in mass from about 0.08 to 100 solar masses, although the majority of stars have masses similar to or less than that of the Sun. Masses greater than 10 solar masses are rare. However, the range of densities is very great. Red giants, such as Betelgeuse, are mainly less dense than the air we breathe, whereas a sugar-lump size of white dwarf material would, on Earth, weigh in excess of 1 tonne.

The main sequence extends from hot stars of high luminosity in the upper left corner of the H–R diagram, to cool stars of low luminosity in the lower right corner. White dwarfs lie sparsely scattered in the lower left corner, and red giant stars of great luminosity and size form a horizontal band that joins the main sequence near the middle of the diagonal band. Above the giant stars, there is another sparse horizontal band where the super-giant stars live. The stars in the lower right corner of the main sequence are the red dwarfs, and the stars between the main sequence and the giant branch are called subgiants. The significance of the H–R diagram is that stars are concentrated in certain distinct regions instead of being distributed at random, a clear indicator to those who first constructed it that there are laws that govern stellar structure and evolution.

A star's life depends on its mass: the more massive it is the shorter its life. A star's life can end in several ways, but all end because the hydrogen starts to run out.

Stars whose main sequence spectral class was anywhere from M on up through the As will start its endgame by slowly expanding into a red giant. Because nuclear fuel is no longer plentiful in the core, it cannot support itself and it begins to contract under its own weight. As it collapses it gets hotter. Soon, the layer immediately above the core will become hot enough to fuse hydrogen into helium and the layer will have an ample supply of hydrogen (unlike the exhausted core). It becomes a hydrogen-burning shell, and will actually burn hydrogen into helium faster than the core did during its main sequence lifetime.

The added energy and outward pressure from this hydrogen-burning shell stop the collapse of the upper layers, in fact reversing the contraction, which will keep expanding until the star becomes a red giant. But after a few million years, the new hydrogen-burning shell will exhaust itself, also causing the star to contract under its own weight once again.

Briefly, the super-compacted core may flash into life, fusing helium into carbon for a time measured literally in seconds. However, since helium fusion produces much less energy than hydrogen fusion does, and since the core is buried so deeply within the star, this helium flash will not be seen. Soon the star sheds up to half its mass, which expands as a shell of gas around the troubled star into interstellar space.

The core that the gas leaves behind is a fascinating object, as we saw when we considered Sirius-B. It has the mass of a star but the size of a planet, it is a cooling cinder with no internal heat source of its own. This white dwarf will now do nothing but cool.

The small surface area and large heat capacity of a white dwarf mean that it takes a long time to cool off – longer, even, than the currently estimated age of the Universe. Someday in the far-distant future, much of the expanded and thinned-out Universe will consist of islands of white dwarfs that have cooled to become benign, black worlds, forever drifting through the darkness of a Universe that once remembered stars. But becoming a white dwarf is not the fate of heavier stars.

A class B main sequence star will leave the main sequence much as lighter stars do, collapsing a little, forming a hydrogen-burning shell, turning into a red giant, shrinking again as its hydrogen-burning shell exhausts itself, then shining more brightly as its core goes through a helium-burning phase. The difference is that burning helium into carbon in the star's core is no longer the end of the road. As this fuel supply runs out, the star's collapse reignites the depleted hydrogen-burning shell and turns it into a helium-burning one, with a new hydrogen-burning shell in a layer above the old one and the core underneath. In the core it will be hot enough for carbon to be fused together with helium to form oxygen.

When it runs out of carbon the star reorganises itself again, fusing oxygen into neon, the old helium-burning shell can become a carbon-burning shell, the formerly outermost hydrogen-burning shell becomes the new helium-burning shell, and yet another thin hydrogen-burning shell emerges outside of that. Then neon can fuse into magnesium, then magnesium can fuse into silicon, and so on down the periodic chart until, finally, chromium gets fused into iron.

But all this fusion burning of successively heavier elements is forlorn. The star is doomed: each tier of fusion produces less energy than the preceding stage and the star rapidly exhausts its nuclear fuel. During these late stages it can become bloated to hundreds of times the diameter of the Sun, becoming a red supergiant like the beautiful Betelgeuse.

When the core starts to fuse iron it becomes cooler, not hotter, and the outward, supporting pressure that its core had been generating disappears. The star's core collapses in the blink of an eye and becomes a neutron star with a surface gravity of a million million times that of Earth. But that is not the end of the matter by any means. As it collapses, the star liberates a lot of gravitational energy and the outer layers of the star become superheated plasmas hot enough to fuse their constituent ions into not only iron, but copper, strontium, silver, gold, lead and even uranium. These super-hot, super-bright outer layers are violently ejected into space carrying their newly formed heavy elements with them and creating one of the most spectacular and rare sights in the Universe – a supernova. Through our telescopes a supernova can be seen as an expanding nebula, the wreckage of a once bright and mighty star.

The end of the real heavyweight stars, such as spectral type O, begin their final phases as the middleweight stars do, bloated with energy-producing shells around the core, fusing heavier and heavier elements together until they reach iron. Once again a catastrophic collapse occurs, but what is formed may not be a neutron star. There is an upper limit to the mass of a neutron star and if the star's core exceeds it then no known force can support its weight and it becomes a black hole.

Not far from our Sun is the star Beta Hydri, only 24.4 light years away. It is easily visible from the southern hemisphere and is the nearest conspicuous star to the unmarked south celestial pole, around which the southern sky turns. It is a yellow-orange star of about 1.1 times the mass of our Sun and 1.5 times its diameter. It is brighter than our Sun, being 3.5 times its luminosity. Why is this star of interest? Because it represents our Sun's future. It is 6.7 billion years old and is a Sun-like star that is coming to the end of its hydrogen-core burning phase and is evolving away from the main sequence.

Beta Hydri's surface abundance of lithium is far greater than our Sun's. Lithium has been made in its core and at its age its convection zone reaches deeper into the star and dredges up this and other elements that would be hidden in a younger star. It is a star that is slowly dying. Its transition region between its chromosphere and its corona is almost gone and it seems well on the way to losing its corona. Monitoring of the star suggests a substellar companion, possibly a gas giant planet, in a close 45-day orbit.

If there were an Earth-like planet at an Earth-like distance from this star, it would have been scorched and rendered uninhabitable. Only at a distance of 1.9 times the Earth–Sun distance would any planet orbiting Beta Hydri be warm enough to have water and cool enough to keep it. As we shall see, there are ways to survive the evolution of a Sun-like star. Perhaps the putative residents of the Beta Hydri star system have left and moved on to another star, like Delta Pavonis, which is 9.3 light years away from Beta Hydri. They might find some respite there but only for a while, because Delta Pavonis is also older than our Sun.

There is an interesting science fiction novel written in 1959 by the late astronomer Sir Fred Hoyle called *Ossian's Ride*. It deals with aliens coming to Earth, fleeing the disaster of their star becoming a red giant.

The Kamiokande detector looks for particles from the heart of the Sun.

CHAPTER TWENTY-SEVEN

End of days

The Aztecs foretold a time 'when the Earth had become tired ... when the seed of the earth has ended'. In the poem 'Hymn to Intellectual Beauty', Percy B. Shelley writes:

The aweful shadow of some unseen Power
Floats though unseen amongst us, – visiting
These various Worlds with as inconstant wing.

Our Sun is 4.6 billion years old and it has already burned half of its hydrogen fuel at its core. The foreseeable generations of mankind are safe – there will not be any fundamental changes in the Sun for them and we have hundreds of millions of years of a steady Sun ahead of us.

But someday it will change. There will be a last perfect day on Earth. The Sun will cease to be our friend and the giver of life. In Isaac Asimov's famous Foundation novels, humans have populated the galaxy and have only dim memories of their home planet and star. The green fields of Earth will be no more – what the Sun has given the Sun will take away. If mankind, or whatever we become, survives, and if we are still around here, we will have to cope with a swelling Sun. However, remarkably, we may already have a solution.

To find out what is in store we must look to the stars. The object astronomers call NGC 7027 is a star entering the final stages of life, going through spectacular death throes as it evolves into what they call a 'planetary nebula', a cloud of gas. The term planetary nebula is a misnomer and came about not because of any real association with planets, but because in early telescopes objects of this type often appeared planet-like, even glowing with a bright green colour. We now know that green colour to be from very hot oxygen atoms in the gas surrounding the central star.

What has happened to the star that became NGC 7027 is that after it used up most of its nuclear fuel, profound changes occurred in the star, not all entirely understood. It began to pulse and the pulsation combined with the pressure of light from the star drove the outer layers of its atmosphere away, forming an extended expanding envelope around it, so large that, when it happens to our Sun, the gas expelled might extend many times farther out than Pluto. During this period the star loses material at very high rates; a star several times the mass of the Sun might shed an amount equal to the total mass of the Sun in less than 10,000 years.

Observing NGC 7027 and objects like it, we see that the gas in the circumstellar envelope is mostly made up of simple molecules such as molecular hydrogen and carbon monoxide, combined with other gases such as cyanide, sodium chloride and possibly water vapour. Complex hydrocarbon molecules are also known to be present in circumstellar envelopes. Significantly, the material cast off includes the elements that are key to the origin of life as we know it – carbon, nitrogen and oxygen – elements that are created by nuclear fusion of hydrogen and helium in the dying star's core.

While the colourful envelope is being formed and ejected into interstellar space, things are changing on the central star. Its surface heats to temperatures in excess of 200,000 degrees Celsius, giving off prodigious amounts of high-energy radiation. The increase in ultraviolet radiation first rips apart the molecules in the envelope and ionises the constituent atoms. But this phase is short, perhaps less than 1000 years. This means that we have caught NGC 7027 at a very important time, right in the middle of this transformation. The molecules in the patterned envelope of

NGC 7027 are being destroyed. NGC 7027 is an object that will survive in its present state for only the blink of an eye, in galactic time. Such a fate could be our Sun's.

Our Sun will reside peacefully on the main sequence for 11 of the 12 billion years of its life. But that is not to say that there will not be changes; there will, and they will affect the Earth and mankind profoundly. Some of what we think is in the future for our Sun is conjecture, some is based on observations of similar stars that are going through these phases. Many ideas are gleaned from the sophisticated computer simulations of what will happen to the Sun as it ages and slowly runs through its finite supply of nuclear fuel.

For the next several billion years the temperature of the Sun's surface, and consequently its brightness, will increase, being about 10% up in the next 1.1 billion years. Some believe that as the brightness of the Sun increases the concentration of water vapour in the Earth's atmosphere will increase as well, rapidly leading to the possibility of a runaway greenhouse effect on our planet that could turn the Earth into another Venus.

Stars like our Sun but younger are thought to be dominated by sunspots, but in our Sun it is the bright regions called faculae – bright regions above the photosphere – that are dominant, becoming increasingly so, we believe, as the Sun ages. Faculae emit far more ultraviolet than visible radiation, which would create more ozone in the Earth's atmosphere.

With the increased brightness of the Sun the ice caps will melt and the sea level will rise. Being warmer, the Earth will have increased rates of rainfall and evaporation and an altogether stronger water cycle as well as stronger winds, all of which will contribute to enhanced erosion. Some calculations suggest that as a result of this in 900 million years the amount of carbon dioxide in our atmosphere will have fallen to a level where plants will have problems surviving. If we lose the plants we will be in deep, possibly terminal, trouble. Within the next billion years the enhanced ultraviolet radiation could destroy the stratosphere and evaporate the oceans. The Earth could be an inhospitable, uninhabitable wasteland long before the Sun dies.

It was thought that when solar-type stars started to run out of nuclear fuel they would expand to several hundred times their original diameter, swallowing up any nearby retinue of planets. As we have seen the swelling is a reaction of the outer layers to significant changes going on deep in the star. The core – the powerhouse of the star – contracts significantly after the hydrogen fuel has run out, and when that happens a solar-type star undergoes two phases of expansion. The first one ends with the core being compressed to a higher temperature than it had during the steady phases of its life, high enough for helium fusion to take place and, for a while at least, some semblance of stability to be re-established. The stellar core goes into a less condensed state and the outer layers become less expanded.

But the helium fuel the star is consuming to maintain itself will not last long, it is drawing on its rapidly diminishing reserves. After the stellar core has consumed its helium it must contract yet again, which leads to a second expansion of the outer layers and the star becomes a supergiant, growing much larger and much more luminous than the original star used to be. The structure of the star is now more complicated. The helium at the core runs out and the star's energy comes mostly from a hydrogen-burning shell around a dense core of nuclear ashes. As the core becomes denser and more massive with time, rising temperature and pressure in the hydrogen burning shell increase the energy output. Our Sun will never burn so brightly as it does towards the end.

How would the Earth resist the brightening Sun? We know that in the past the Earth responded to changes in the Sun's output, adjusting itself so that over billions of years a roughly constant set of conditions existed on its surface, allowing life to develop. It seems that in the past the Earth had more carbon dioxide in the atmosphere, which gave it more of a greenhouse effect, and the greenhouse effect declined at the right rate to keep pace with the brightening Sun so that the Earth's environment was relatively stable. Some call this the Gaia hypothesis – that the Earth will act as a self-regulating system to maintain the conditions for life to exist. Personally I think that the Gaia idea reads too much anthropomorphical purpose into a nonlinear, self-regulating system that has been seen to be stable between certain limits in the past. But whatever your thoughts about the status of the Gaia hypothesis, it will not rescue us in the future. The changes that are coming will be beyond the Earth's ability to cope and any life forms that may want to outlast them will not be able to rely on a naturally adapting biosphere for protection. They will be on their own.

So the fate of our Sun is sealed. In about 7.5 billion years' time the expanded Sun's luminosity will peak at several thousand times what it is today. Then, when too little envelope mass is available to feed the hydrogen-shell burning zone, the outer layer will be puffed off like NGC 7027 and a white dwarf star will remain to cool, almost forever. It once seemed that the destruction of the Earth was inevitable in the face of such cataclysmic changes in our source of light and warmth, because the Earth would have been fried and then swallowed by the expanding Sun, but now astronomers are not so sure.

Detailed calculations suggest that in the later stages of its life the Sun will lose mass and increase its size to a radius of 168 million km, much larger than the 150 million km distance at which the Earth orbits the Sun today. No planet can survive for very long when touched by the expanding envelope of its encroaching star: the drag on its orbital motion would cause the planet's path to decay and it would spiral inward and disintegrate. Because of this Mercury and Venus are doomed. Venus is an example of what could have happened to the Earth; it is the same size. As the Sun loses mass its gravity is lessened. Because of this Venus's orbit will expand from 108 million km to 134 million km, but that is not enough to save it. Once inside the photosphere the orbit will decay rapidly due to drag. A few thousand years at most and the planet will fall to the centre of the Sun, fragmenting as it does so to scatter itself across the face of the star that created it all those billions of years ago.

This was thought to be the fate of the Earth. However, possibly the Earth has been granted a reprieve and we might not be swallowed up when the Sun dies in about 7.5 billion years' time. The new calculations actually extend the length of time the Earth will be habitable by 200 million years. Nevertheless, in the end the surface of the planet will simply become too hot for life to survive. Earth dwellers will have to find alternative homes. According to new calculations, the orbit of the Earth will increase slightly beyond the outer atmosphere of the red giant, as its gravitational pull weakens. The key point is that as the Sun loses mass the radius of the Earth's orbit will expand because of the smaller mass that is available to keep it in orbit. The latest estimates are that it will increase to 185 million km, just enough to save it.

But what would be worth saving? The Earth will look like Mercury, a wrecked, baked and blasted, bone-dry, scarred hulk with the exposed floors of former oceans. From the Earth at this time the leering red Sun would cover 70% of the sky, as the Earth would be orbiting just 10% of the solar radius away from it.

Further out in our reshaped solar system the orbit of Mars will expand, saving it from destruction, as their expanded orbits will save Jupiter, Saturn, Uranus, Neptune and Pluto. As for the material in the planetary nebula puffed out when the Sun goes through its convulsions of old age, it will have little impact on the planets themselves. The outer layers of a red giant are extremely tenuous, by terrestrial standards a fairly decent vacuum!

This is a long time in the future and when it happens mankind, or what it is to become, may have long ago left the planet or died out. Perhaps our future world will be devoid of life or perhaps, with mankind absent, evolution will find new pathways and designs, as we know it must. Perhaps the threatening and encroaching Sun will slowly fry a planet of strange but not intelligent creatures. Or perhaps it will be witnessed by new intelligent lifeforms, nonhuman, who have evolved after our kind has changed, died out or departed.

However, supposing anyone or thing cared, how soon would the Earth be uninhabitable? Could we planet hop outwards Goldilocks like, taking advantage of the expanding zone of habitability – not too warm, not too cold – as the Sun brightens? No. We will have to leave Earth before Mars is even warm enough for us. The planets are simply too far apart for planet hopping to be a viable solution.

The warming Sun will render Mars habitable for just 100 million years between 6.1 and 6.2 billion years' time, but given the changes on Earth we would have had to have left by about 5.7 billion years' time. The sad fact is that by the time the Earth is being incinerated, Mars will still be a frozen world and we will not have the time to wait for it to thaw out and become temperate and inhabitable.

A billion years after Mars briefly becomes warm enough for the subsurface ice to melt and turn it into a waterworld with a thicker atmosphere, the wave of habitability will have reached the hitherto frozen Europa, moon of Jupiter. Today Europa is a world with a frozen surface and probably a sub-ice ocean beneath its ice crust. Perhaps life has evolved there? Europa is certainly the focus for much speculation at the moment and will probably be visited by space probes before two decades have passed. But if it has life at the moment, or will have in the future, the brightening of the Sun will melt the ice crust, turning it into an ocean Moon. Perhaps mankind could cling on there in great floating or submerged cities? Is Europa a possible outpost for the human diaspora? Again, no. Europa would be habitable but briefly, for less than 100 million years, and where would we go in the billion years between Mars becoming nasty and Europa becoming possible?

It is the same story as we travel outwards in the solar system. Some of the moons of the gas giants Jupiter and Saturn could well become places where we could migrate to as our habitations became baked, but the timing would be all wrong. There would be no overlap between them. After Europa turned hostile we would have to wait 50 million years for Titan, Saturn's major moon, to be available for a brief period. Oberon would be available for a few million years, but Triton, Neptune's major moon, and Pluto never would.

Saving the Earth

However, perhaps there is something that could be done to save the Earth. Perhaps we could move it out of harm's way? This is not such a far-fetched solution. After all we can only speculate about the powers and energies that our far distant ancestors may have available to them. Remarkably, even in its current stage of development mankind will soon have the ability to move the Earth into a new orbit, a manoeuvre that may more than double the time we can survive on our planet.

It all depends on using the well-understood 'gravitational sling shot' technique that has been employed to send space probes to the outer planets. A large asteroid could be used to reposition the Earth to maintain a benign global climate. It could ensure humanity's survival for longer than it could last if we could not move the planets. In fact, it could allow our descendants to redesign our solar system, moving moons and planets into orbits so that we can hop from new Earth to new Earth.

What is required is for a large asteroid, about 100 km across, to fly past the Earth, transferring some of its orbital energy to our planet whose orbit would expand slightly. The asteroid would then move out to encounter Jupiter, where it would acquire more energy that it could impart to the Earth on a subsequent pass. The beauty of the technique is that we have plenty of time to carry it out. Initially we would have millions of years to select the appropriate asteroid and develop the necessary technology to deflect the giant rock in the direction of Earth. Simple calculations suggest that to expand the Earth's orbit around the Sun at a rate that compensates for the increasing brightness of the star would require an asteroid encounter every 6000 years, or about every 240 generations.

Earth's gradual outward migration may require adjustments to be made to the orbits of other planets as well. Recent calculations of the solar system's stability indicate that if the Earth were removed then Venus and Mercury would become destabilised in a relatively short time. It would be a procedure that required some care. If the 100 km asteroid were to collide with the Earth then it would wipe out all life on our planet.

After all this has happened the Sun will become a white dwarf. The sphere of gas it ejected will drift off and eventually be gathered up in a new cloud, to become part of the next generation of star formation. Perhaps one day the scattered envelope of the Sun will throw its lot in with another star to be born, live, die and, perhaps, give sustenance to other warm little planets.

There are many white dwarfs in our galaxy and perhaps, as novelist Isaac Asimov predicted, we will move out into the galaxy and lose contact with our birth star and there may come a time when we no longer know where it is. Eventually the stars in our galaxy will start to die and it will fade. Stars will explode at an ever slower rate and the growing number of white dwarfs will slowly leak into space, our Sun among them, to begin an endless solitary journey through the eternal darkness. Such is the fate of the star that gave us so many happy, sunny days.

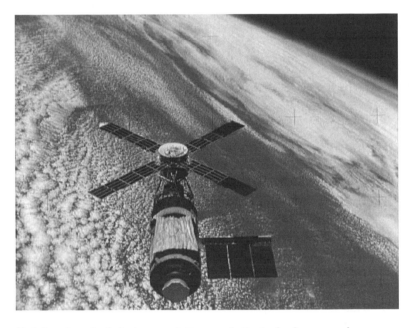

Skylab – America's first space station revolutionised solar research.

CHAPTER TWENTY-EIGHT

The wind from the Sun

The energetic solar storms that pummelled the Earth in the autumn of 2003 had an effect that was felt to the edge of the solar system, causing disruptions on other planets as well as the Earth. In a three-week period between October and November more than a dozen storms, including the most powerful ever measured, erupted from the face of the Sun, sending blast waves and streams of particles into space. The storms caused little damage on Earth, mainly because most of the very powerful blasts were not aimed directly at our planet and because we have grown accustomed to taking effective precautions. There were spectacular displays of aurorae, many seen at low latitudes. There were some radiation effects that required some rerouting of aircraft travelling on polar routes, some minor disruption of some satellites, and astronauts aboard the International Space Station had to seek temporary shelter in the more shielded parts of the station.

Billions of tons of plasma from the blasts flew through space at speeds of up to 10 million km per hour, the fastest ever measured from the Sun, and as the blast wave expanded through the solar system sensors on numerous interplanetary space probes recorded the disturbance as it passed by.

One blast damaged the radiation monitor aboard the *Odyssey* spacecraft orbiting Mars. However, it was able to record how the blast wave stretched and tore the thin Martian atmosphere, stripping part of it away. Scientists said that the effect could partly explain how Mars has lost so much of its atmosphere and water over the past 3.5 billion years. It could be that rare, violent storms from the Sun, so infrequent on a human timescale, could add up over billions of years to turn Mars from the Earth-like world it obviously was, with a thicker atmosphere capable of supporting rain and running water on its surface, into the barren, arid world it is today.

The blast wave continued beyond Mars and headed towards the outer planets, disrupting the magnetic field around Jupiter and setting off a week-long burst of radio emissions that were detected by the *Ulysses* spacecraft in its wide orbit around the Sun monitoring its solar wind. The shock front caused a similar event when it arrived at Saturn, and the *Cassini* craft recorded it as it approached the planet.

As they travelled further out the blast waves merged, creating a front that is headed towards the edge of the solar system. Six months after they set off the waves reached the *Voyager 2* spacecraft, which is 11 billion km from the Sun, and a few weeks later they touched *Voyager 1*, 14.5 billion km distant. *Voyager 1* was launched in September 1977 and *Voyager 2* in August 1977 on a somewhat slower trajectory into the outer solar system. They are the grandest voyages of exploration ever undertaken. *Voyagers 1* and *2* reached Jupiter in 1979 and *Voyager 1* reached Saturn in 1980, with *Voyager 2* reaching the ringed planet the following year. Saturn was the last planetary encounter for *Voyager 1*, but *Voyager 2* went on to Uranus and Neptune in 1986 and 1989. Then they ceased being planetary exploration missions and became interstellar ones.

By the end of 2004 the merged blast wave from the 2003 flares is expected to reach the edge of the solar system, about 5 billion km further out. There it will encounter the heliopause – the end of our Sun's influence and the start of interstellar space. This is the bubble in space produced by the solar wind. Virtually all of the material in the heliosphere emanates from the Sun itself. Scientists expect the blast wave to temporarily push out the heliopause some 400 million miles when it hits, with it rebounding to its normal position in a year or two. The heliopause is our final destination in our journey from the centre of the Sun. It is the end of the Sun's empire.

Scientists hope that *Voyager 1* will cross the heliopause. At some distance from the Sun, well beyond the orbit of Pluto, this solar wind must slow down to meet the gases in the interstellar medium. It must first pass through a shock, the termination shock, to become subsonic. It then slows down and gets turned in the direction of the ambient flow of the interstellar medium to form a comet-like tail behind the Sun. Currently there is some controversy about whether *Voyager 1* has reached the termination shock or even passed into the heliopause. One team of researchers at the Bell Laboratories and the New Jersey Institute of Technology say they have evidence that *Voyager 1* is actually in the vicinity of the termination shock and had even passed briefly into the heliopause. Other scientists are unconvinced.

Of all the objects mankind has ever launched into space, only four will leave our solar system: *Pioneers 10* and *11* and *Voyagers 1* and *2*. On the side of each of them is a message from the creatures who built it. The Pioneer plaque and the Voyager interstellar record have been designed so that they can be read with the minimum of information, requiring knowledge of the galaxy, physics and binary numbers. It is a reasonable assumption that any aliens who might find it have this much knowledge. The Voyager record contains images and sounds, whereas the Pioneer plaque is simpler. Both contain a pulsar map and clock.

There are in our galaxy many pulsars – swiftly spinning neutron stars that give off a regular flash like a cosmic lighthouse. The map on board the space probes shows the positions of 14 of them and their time periods. Because the pulsars are slowing down, the map will help any aliens determine where and when the probe was launched. Voyager also has an ultra-pure source of uranium-238 that the aliens could use as a radioactive clock for determining the probe's age as well. The accuracy of the map gets less as time goes by, but it has been estimated that if the space probes are ever found the map should enable its finders to pinpoint the probe's star of origin to within a few hundred stars.

The Voyager record contains 115 images ranging from scientific calibrations, to the planets, people, DNA, the Earth and many other things chosen to be somehow representative of ourselves and the way we see the Universe. Among them there is an image of the solar spectrum – violet to red with the characteristic Fraunhofer lines. There is also a picture of the Sun and an image of a sunset with birds. There is an image of a forest and of a single leaf.

Long after they fall silent, the Voyager twins will keep speeding away from our solar system. Long after our Sun has swelled to become a red giant, the Voyager craft will still be moving among the stars. Perhaps long after mankind itself has disappeared from the cosmos they will still be wandering. If one is ever found by another intelligence, I wonder what they will make of the images of the creatures who made it so long ago and of the star they were born under, now so far away.

The Voyagers could be our final mark on the Universe, what mankind is judged by. If humanity dies out or if we never leave our home solar system, the Voyager spectrum and sunset will be the only surviving pictures of our Sun, showing a star so different from the white dwarf, it is destined to spent most of eternity as.

The Sun rises every 90 mins for Space Shuttle astronauts.

CHAPTER TWENTY-NINE

Coda – the eye of god

The 'winged eye' religious symbol occurs in many cultures across the world. In ancient Egypt it was the eye of Horus, but the Mayans and Aztecs of Mesoamerica and the Nazca of Peru, to name a few, had similar symbols. Sometimes it was called the eye of god. It was our old friend Edward Walter Maunder who believed that the winged sun disk symbol of ancient Egypt, as well as what he called the 'ring with wings' solar disk symbols of Assyria and other ancient Mesopotamian civilisations, were inspired by the Sun's coronal halo seen during total solar eclipses. As well as inspiring ancient Egypt's winged sun disk symbol, this coronal 'bird of the Sun' may have been the inspiration for the phoenix bird of classical Greek myth. There is also an Egyptian myth of a cosmic battle between the solar falcon god Horus and the Sun-eating serpent god Set (or Apop).

The eye of god has also been seen in modern times. On 22 April 1715, Sir Edmund Halley observed a total solar eclipse over southern England and described the sun's corona thus:

A few seconds after the sun was totally covered, there was visible round the moon a luminous ring, in breadth about a tenth part of the moon's diameter. It was of a pale white or rather pearl colour, seeming a little tinged with the colour of the iris, and concentric with the moon.

Numerous Stone Age artefacts representing an eye have been found. Excavations have discovered hundreds of multi-eyed female statuettes in a temple consecrated to the goddess Inanna at Tell Brak in eastern Syria and similar idols have been found at Ur. The Assyrian goddess Mari was reputed to search men's souls with her eyes. In Sumarian religion Ea or Enki was known as the 'Lord of the Sacred Eye'. Apollo was symbolised by the eye and the sun was known as the 'eye of Zeus'. The Hebrews represented the Sun as an eye, and as possessing the attributes of an eye: seeing all, knowing all. In the Bible there is a reference to a city known as En Shemesh, which means 'eye of the sun'. Shakespeare wrote in *King Henry VI* that 'the sun with one eye vieweth all the world'.

The story of our Sun has been told, from its birth in a cloud of gas and dust, to its death as a white dwarf, forever cooling and surrounded by the charred remnants of its inner planets and the chilly gases of its outer ones. In our exploration and under-standing of it we humans have moved from myth to science to grasp all of its moods and complexities.

But its story is also the story of another star. That is a mere 45.7 light years away and can be seen on the northern edge of the constellation Scorpius, just off its left claw. It is a little bigger and brighter than our Sun, at 4.2 billion years old slightly younger and a mere 12 degrees Celsius hotter. It rotates every 23 days, slightly faster than our Sun, and seems to have a 9–11-year sunspot cycle. This is 18 Scorpii – as we have seen before, the nearest the Sun has to a twin in space. Their story is the same for billions and billions of other similar stars scattered throughout the Universe. Our star's life and times tell us a lot about the story of the cosmos.

Arthur Eddington once pointed out that man stands midway in scale between an atom and a star. There are 1 followed by 27 zeros atoms in a human and 1 followed by 28 humans would make up a star. He also pointed out that our lifespan is midway between the excited state of an atom and the lifetime of our Sun. He then went on to refine these estimates, saying that to be more accurate the midpoints of mass and lifespan between an atom and a star should be occupied by a hippopotamus and a butterfly!

The Sun sets again, as we all knew it would. It has once more circled the stone called Intihuatana, which means the 'hitching place of the sun' – an ancient sundial among the remains of Machu Picchu. The Inca, like most peoples, knew that the fate of the Sun and the Earth are linked and that sooner or later something would interfere with its endless cycles. They thought it would be the gods that interfered, but we now know it is science. But science also shows us that man's fate is not joined to that of the star under whose light we were born if we do not wish it to be. Someday we may have to leave the Sun, witness our final sunset and say goodbye. Seen from further along the almost never-ending times that stretch before us, we know that most of the sunsets to come will not make the promise that sunsets make today. Our Sun is our parent, giving us life and a home but we do not live at home with our parents forever.

Is the story of our Sun true? In the nineteenth century the sailors of St Malo in Brittany said that on some evenings, when the air and sea are unusually quiet, if you listen hard you could hear the faint hiss, like a red-hot iron being plunged into water, caused by the Sun plunging into the ocean.

Bibliography

Baxter, W. M. (1963) *The Sun and the Amateur Astronomer*, Lutterworth Press.

Bethe, Hans A. (1991) *The Road from Los Alamos*, Touchstone.

Calder, Nigel (1997) *The Manic Sun*, Pilkington Press.

Chandrasekhar, Subrahmanyan (1983) *Eddington*, Cambridge University Press.

Duncan, David Ewing (1998) *The Calendar*, Fourth Estate.

Eddington, Arthur (1959) *The Internal Constitution of the Stars*, Dover.

Eddington, Sir Arthur (1929) *Stars and Atoms*, Oxford Press.

Galilei, Galileo (1989) *Sidereus Nuncius*, University of Chicago Press.

Gamow, George (1964) *A Star Called the Sun*, Viking Press.

Golub, Leon and Pasachoff, Jay M. (2001) *Nearest Star*, Harvard University Press.

Heppenheimer, T. A. (1984) *The Man-Made Sun*, Little, Brown.

Herman, Robin (1990) *Fusion*, Cambridge University Press.

Jukes, J. D. (1959) *Man Made Sun*, Abelard-Schuman.

Lang, Kenneth R. (2001) *The Cambridge Encyclopedia of the Sun*, Cambridge University Press.

Noyes, Robert (1982) *The Sun, Our Star*, Harvard University Press.

Phillips, Kenneth J. H. (1992) *Guide to the Sun*, Cambridge University Press.

Soon, W. Wei-Hock and Yaskell, S. H. (2003) *The Maunder Minimum and the Variable Sun–Earth Connection*, World Scientific.

Steel, Duncan (1999) *Eclipse*, Headline.

Taylor, Peter O. (1991) *Observing the Sun*, Cambridge University Press.

Zirker, Jack (2002) *Journey from the Center of the Sun*, Princeton University Press.

Index